Excel for Surveyors

Philip Bowcock
and
Natalie Bayfield

A division of Reed Business Information
Estates Gazette, 151 Wardour Street, London W1V 4BN

First published 2000
Second impression 2001

ISBN 0 7282 0332 4

To Sheila and Nonnina

Typeset by Amy Boyle, Rochester, Kent
Printed and bound by Bell and Bain Ltd, Glasgow

Contents

Foreword

In the today's business world the use and manipulation of data is everything.

Last year I was contributing a paper to a seminar and as I was leaving the conference centre, I took the opportunity to peruse the book stall that had been set up in the lobby. There was a good selection of interesting books available on a wide range of property matters but I was searching for something in particular. I needed a book that would help to me unlock the obvious potential that the computer had to assist valuers. I was disappointed. Some textbooks made only a passing reference to PCs, while others referred to spread sheet solutions but assumed that the reader was fully conversant with the techniques needed to explore those solutions.

Valuers are very fortunate since the data and the calculations that they use are well suited to being dealt with by the computer. In particular the opportunity to view a wide range of different scenarios and "What if?" possibilities are an absolute gift to those reluctant to manually compute a large number of alternative calculations. Regrettably however computer literacy has still not permeated to the core of our profession and in particular has not yet fully reached the more senior members of the profession and so this potential is not yet being properly explored.

There are many good handbooks on the use of spreadsheets and Excel, in particular but none of them fully answer the needs of our profession. Many of the built in financial formulae in Excel for example, are directed to American accountancy needs and the terminology does not obviously relate to our specialised expressions.

In order to develop my spreadsheet knowledge I approached Philip Bowcock. We had worked together on unusual valuation problems in the past and I knew that he had the combination of valuation experience and computer knowledge that I needed. It was in fact from our conversations together that the idea arose of producing a book that would help a wide range of valuers, both those that have no Excel knowledge and also those that have some, but not quite enough to take full advantage of the computer's power.

*Excel for Surveyor*s, the book that has now been produced by Philip Bowcock and Natalie Bayfield is the book that I have long sought and there will be few valuers and surveyors who cannot benefit enormously from its simple and clearly set out explanations. Not only can it be used to learn how to exploit Excel, it will also stimulate ideas to look at valuation calculations in a new and creative way.

Anthony Salata
Jorden Salata Graham
Former Chairman RICS Dispute Resolution Practice Panel
St James's, London

Preface

Philip Bowcock recently retired from the post of Lecturer in Valuation at the University of Reading after a total of 28 years in the Department of Land Management and prior to that in the College of Estate Management before the merger with the University in 1972. He has always been interested in mathematics and has been involved in the use of computers for the past 30 years.

Natalie Bayfield graduated in Property Valuation and Finance at the City University in 1998. She is now a Director of Bayfield Training Ltd, which specialises in teaching Excel to the profession. She has worked with the City University, the University of Ulster, the University of Westminster and the Investment Property Forum, instructing on the use of computers for property valuations.

This volume is intended to introduce surveyors who may not have had the benefit of a course of study which included information technology and the use of computers. It does not set out or recommend any particular format or method for making valuations.

The formal name of the software discussed is Microsoft Excel™, but for convenience we shall refer to this simply as "Excel". Screen grabs are reproduced by kind permission of The Microsoft Corporation.

The text was produced on a Toshiba PC using Windows and Office 97 (Natalie) and a Macintosh 7600 using Office 98 (Philip).[1] We found few practical differences between these versions and Version 5 of Excel. Screen shots were also produced on the Macintosh and edited where necessary with Adobe Photoshop 5.0™.

Much of the content of this volume can be used with earlier versions of Excel, and it should be possible to set up many of the earlier examples with no modification.

We are very grateful to the staff of the Estates Gazette and to Colin Greasby, in particular, for their encouragement in proceeding with this work.

Philip Bowcock and Natalie Bayfield

---

[1] The 97 version is for PC and the 98 version is for Macintosh. For all practical purposes they are identical.

Chapter 1

Your computer

There can be very few offices today which do not have among their equipment a modern computer. It is well known that developments are moving so fast that today's "state of the art" machine will be considered obsolete in about three year's time, but for the purposes of this volume the facilities offered by machines currently available should be adequate for the foreseeable future.

This is not intended to be an introductory textbook on how to use a computer. However, for those who have a machine but maybe have never switched it on, or having done so are at a loss to know what to do next, we give a short explanation.

The object of the first three chapters is to explain the use of some of these facilities to practising surveyors and valuers, many of whom were at work before modern computers were thought of. It is not easy to come to terms with machinery which is internally very complicated, but which schoolchildren are using every day.

Today there are two main types of desktop computer (which term also includes laptop computers). The most common one is the PC (personal computer) using the system developed by IBM and now incorporated into Microsoft Windows®. The less common one is the Apple Macintosh (generally agreed to be easier to use). Microsoft Excel will run on either system with no significant difference. Files are also interchangeable between the two – if you are using a Macintosh just copy files to a PC-formatted floppy disk and put it in the drive of the PC.

There are two principles to be considered when acquiring computer hardware and software – broadly you get what you pay for, and there will be something bigger, better and cheaper available tomorrow! Over the last 15 years the prices of computers have consistently declined in real terms regardless of inflation and the facilities which they provide. If you are currently considering purchasing a computer for use with Excel then any of the current IBM variety or Apple Macintosh machines are almost certain to have adequate memory and file space.[2] However if you intend to store graphics, such as photographs taken with digital cameras, with your data then you may need substantially more memory and disk space. Presentation software such as Powerpoint, which is part of the Microsoft® Office suite, may also require more memory

There are numerous pieces of information in a computer specification, but the two most significant are the "memory" or "RAM" and the "disk" or "filestore". "Memory" is the working part of the computer where all the information is held and the calculations are done while you are actually using it. It can be thought of as your desk, on which you can work on your document. Internally it consists of many millions of tiny electronic switches which are organised to hold the data that you are working on and translate it into a form in which you can view it on your

---

[2] By "current" we mean any PC Pentium or Macintosh Power PC machine.

monitor. As soon as the machine is switched off all contents of the memory disappear for ever (like sweeping all the paper on your desk into the bin).

"File space" by contrast may be thought of as a bookshelf on which you can put away your work when you have finished. In the computer this is primarily the "hard disk" which consists of one or more rotating metal disks with a magnetic surface in a sealed unit on which data can be stored in a similar manner to the familiar tape recorder. The difference is that the recording head is not in contact with the disk but is very close indeed to the surface – a distance much less than the diameter of the human hair. Consequently hard disks are fragile and must never be opened except in a "clean" workshop.

The "unit of measurement" for both these spaces is the "byte" which consists of eight on-off electronic switches known as "bits". A set of eight bits has 256 possible on-off combinations, each of which can represent one numeric or alphabetic character, a punctuation mark, or other symbol, or a special code for the computer. These are known as the ASCII codes (American Standard Code for Information Interchange).

You should expect to have at least 32 megabytes (million bytes) of memory and 2 gigabytes (thousand million bytes) of disk space.

The monitor is the means by which the computer communicates with you, and the most common size for desktop monitors is 15 inches. The next size, 17 inches costs a little more but can be well worth it by enabling you to see more of the worksheet. Portable computers are of course limited by the size of the case, but most can be linked to an external monitor if desired. Do not forget to observe the general advice on posture when using the computer for long periods. The Health and Safety (Display Screen Equipment) Regulations 1992 relate to use of computers in places of employment but should be used as guidelines all situations.

Every computer requires an operating system to make it work. This is loaded up at startup and produces the pictures on the screen, maintains the file system and many other functions. Current versions (Windows 98 and Macintosh 8.0 upwards) operate as we shall describe. If you use earlier versions you may find slight differences in operating Excel.

It is impossible to discuss every possibility in considering software as complicated and versatile as Excel. Having described the principal operations likely to be of use to surveyors, we suggest that you work through each example we have described in order to become familiar with it, but then try out the other options for yourself.

It is hoped that this text will enlighten those who are still somewhat diffident; suggest new possibilities to those who have used some of the basic operations, and generally remove some of the mystique which still discourages many surveyors.

Finally, you should remember that using Excel efficiently, like using many other complicated tools, requires practice and regular use. Don't expect to remember everything immediately, and be prepared to revise your knowledge from time to time.

Chapter 2

Introduction to Excel

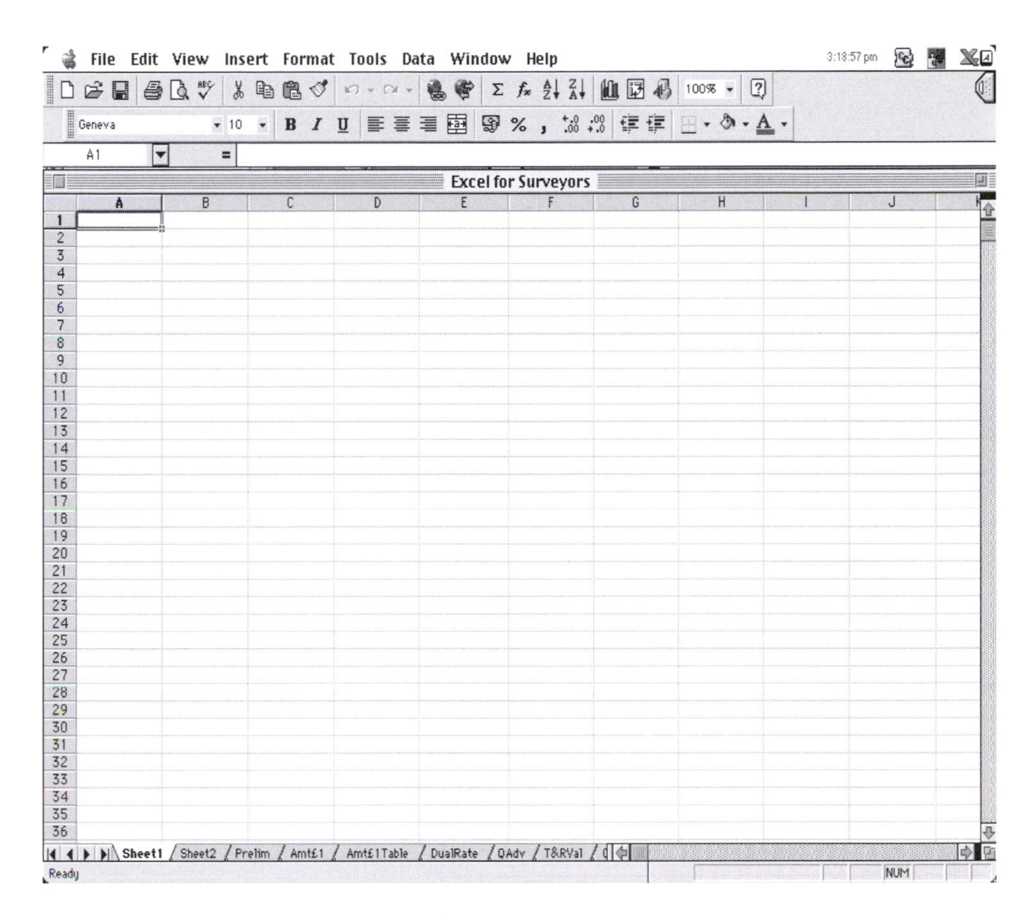

2.1 Description

Microsoft Excel is the most widely used worksheet for desktop computers and has been around since version 1 was marketed in 1985. The very first worksheet program, VisiCalc[3] (1978), was followed by another, Microsoft Multiplan, in 1983. These pioneers were remarkable for their facilities in their time but by comparison with the current version they were very limited in the operations they could perform. Today, Excel provides such a large variety of resources that it is most unlikely that anyone will use every single facility.

---

[3] VisiCalc was conceived by a student, Dan Bricklin, who became frustrated at using paper, pencil and calculator, or alternatively programming a large computer, to solve modest-sized mathematical problems. He decided that a computer would be more efficient than a blackboard for conveying information to students.

There are other worksheets around, notably Lotus 1–2–3 and Clarisworks, but many of the features are similar, and you should be able to carry out many of the operations discussed later with these other applications.

This chapter is designed to take you through all the common features of Excel, the commands and language that will enable you to move around Excel, and some of its little tricks, so that you don't have to refer to other resources. However there are no large examples since this is just a tour. Some of the details may seem tedious if you have used Excel before. If it still seems complicated after you have finished this chapter, don't worry. Once you put it into practice on valuations in later chapters everything here will become second nature.

If you are accustomed to using word processing software, the first thing you will notice when you open Excel is that your monitor shows not a blank page, but a grid of cells. Word processing software, such as Word, assumes you want to fill your screen with text whereas Excel is based on an array of rows and columns. (If your screen is blank when you load Excel take your mouse-operated arrow to the word File at the top left corner of the screen, click once with the left mouse button[4] and then click again on the word New.)

2.2 The mouse

You will notice there is a marker which is called the "cursor" on the screen and as you move your mouse around on the desk the cursor moves around on the screen in the same direction. You will also see that as it moves it sometimes changes shape – around the sides of the screen it is an arrow, in the fourth line down it is an I bar, and when over any part of the grid it becomes a thick white cross. This is the selection cross.

"Clicking" means hovering the cursor over a cell or another part of the screen which you want to select and pressing the left mouse button once unless stated otherwise, eg click twice or click with the right mouse button. If you click a cell its border becomes black and the cell is then "selected".

"Drag" means select a cell, hold down the mouse button, and move the pointer to another cell. The first and last cells and all those in the rectangle formed will be selected – try it. This is very useful if you want to carry out an operation on more than one cell, for example to clear all data or change the appearance of the cells.

Above the cells you will see a horizontal grey strip with the column headers labelled A, B, C, etc. To select an entire column, click on the column header. The entire row or column will appear black except for the top cell. Try this.

Similarly, in the vertical grey strip to the left of the cells you will see the row headers labelled numerically 1, 2, 3, etc. Rows can be selected in the same way.

---

[4] The Macintosh computer normally only has one button and you click once or twice depending on the operation. If one click does not work try two. Double clicking does not usually cause any problem if you do it accidentally. (The word "clicking" comes from the sound made by the buttons on the mouse when they are pressed.)

2.3 Cell references

Each cell in the worksheet has an address or cell reference rather like an Ordnance Survey sheet or map. Each cell is in both one column and one row at the same time. The letter at the top of that column and number at the beginning of that row together define the cell's address or reference. For example when the black border is in the first column on your grid and the first row, you have selected cell A1. Excel also displays the cell reference in the name box towards the top left corner of the screen. Select a cell and see if you can find its reference displayed in the name box.

Worksheets can be huge. Each worksheet in Excel 97/98 can have 65,536 rows and 256 columns. Within this, the limitation is the total amount of memory available in your machine. Quite what one would do with such a large worksheet is difficult to imagine, but no doubt somebody somewhere has tried. In practice it is usually much better to separate parts of a very large project into several different worksheets within the same workbook.

2.4 Some conventions

Throughout this volume we shall use the following:

Return means press the Return key.

Control means hold down the Control key while you press the indicated character key. The ALT key, and on the Macintosh the Command and Option keys are used similarly

Following common practice, words indicating a keystroke are printed in this style of text.

2.5 Worksheets and workbooks

Excel assumes that you want to fill your screen with numbers, and the grid provides these handy little cells in which to put them. This is because each individual number in your worksheet is a separate item. Most importantly, by separating numbers in this way we give them their own unique home and address on the worksheet, which is important when we come to writing formulae. Excel, incidentally, refers to its grids as "Worksheets", but elsewhere you will hear them referred to as "Spreadsheets".

The grid which first appears on your monitor is one worksheet in a workbook which usually contains several worksheets. These worksheets appear to be stacked behind one another. If you look to the bottom of your current worksheet you will see a row of tabs labelled sheet 1, sheet 2, sheet 3 etc. The current sheet tab is coloured white and the others grey. If you click once with the left mouse button on each of the tabs you will swap between the different sheets in your workbook.

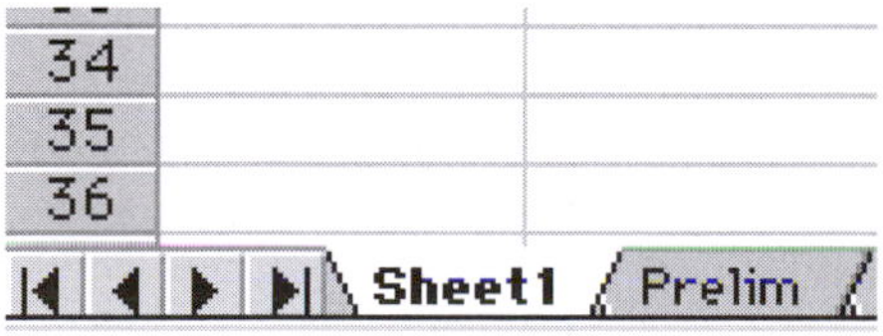

If you click twice you can rename each sheet to something more appropriate to the work you are doing.

2.6 Scroll bars

On the right side and bottom of the worksheet are two grey stripes with an arrow at each end and a darker grey square somewhere between. These enable you to move around your worksheet to areas which are not in view. You can either click on one of the arrows at either end or click on the rectangle, hold down the button, and then drag it one way or the other.

2.7 Entering numbers

To enter a number into the worksheet you need to first tell the computer where to put it by selecting a cell. Right now a cell will already be selected on your worksheet. The selected cell is the one with the thick black marker[5] around its border. To change the selection click on another cell and the thick border will jump to this new position. Alternatively you can use the arrow keys on the bottom right hand corner of your keyboard to move to another cell – try it.

Once you have selected a cell you can enter into it any number that you wish. Select a cell and type in the number 10.

While the cursor is still flashing next to your number 10 you are in editing mode and the computer thinks that you still haven't finished. Perhaps you really intend to put 100.

Press the **Return** key now.

The black border will move to the cell below ready for the insertion of another number.[6]

Note that text is normally left-aligned and numbers are right-aligned within the cell. If you enter a number and accidentally include a non-numeric character or space, it will be interpreted as text. Therefore, if intended to enter a number and it is left-aligned you should check that you have not included other characters, including spaces. However it may still be correct if the cell is formatted as left-aligned – see Chapter 4.3.

2.8 What we can do with them

Having entered numbers into the worksheet we will presumably want to do something with them. We can do calculations, adding or multiplying or link them with other data for example, in order to find the solution to a problem. These are performed with formulae and built-in functions which we will discuss later.

We can also change their appearance, underlining, colouring or emboldening for instance; copy them, or move them by the use of commands. Commands also

---

[5] On a Macintosh this is usually Red.

[6] There is an option in Menu ⇒ Tools ⇒ Options ⇒ Edit to change this so that the same cell remains selected.

help us manipulate the workbook, for example by opening, closing and deleting sheets, or resizing cells.

Commands are usually initiated in one of three ways, from the Menu Bar, from the Toolbars, or by a defined combination of keys including the function keys.

2.9 The menu bar

The menu bar is the top line on the screen. Each word on the menu bar heads a menu of commands which can be seen by clicking on the word – hence its name. There are many commands and we shall examine some of them later. When you click on a menu name the full menu drops down and is displayed.

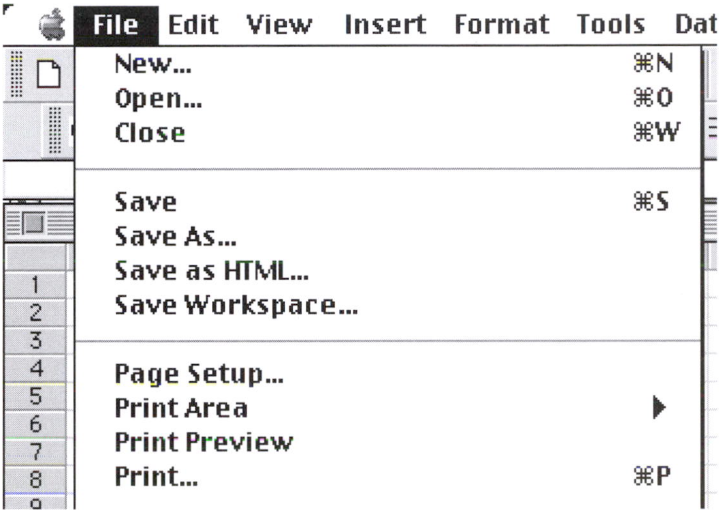

Some commands have a right-pointing arrow next to them. This indicates that there is a submenu of commands which will appear if you follow this arrow with your mouse. This is commonly referred to as "sliding off".

Throughout this volume menu commands will be indicated in the following style. For example

Menu ⇒ Format ⇒ Column ⇒ Width

means move your pointer to the **Menu Bar** and click on **Format**. When the menu drops down slide off Column to Width. In the next box which appears you can enter the desired width of the column(s) in digits. The default for column width is normally 8.43 characters.

2.10 The toolbars

The toolbars appear below the menu bar and provide another way to perform many of the commonly used commands. Normally only the standard and formatting toolbars appear, though there are several others. To see these click on **Menu ⇒ View ⇒ Toolbars** and then in the box adjoining any that you wish to

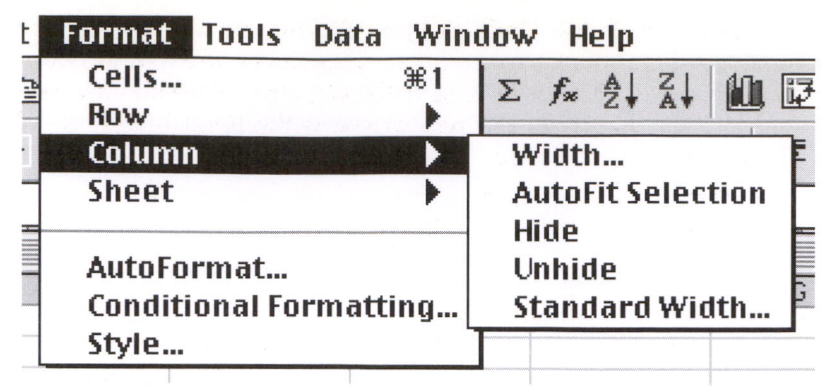

see. In practice these are best left out of view unless actually needed, as they reduce the visible area of the worksheet.[7]

The toolbars show series of icons denoting buttons, which if clicked, perform commands. If you hover over a button with your mouse pointer, without clicking, a label will appear to tell you what command that button performs. For example, to open an existing workbook, instead of clicking on **Menu** ⇒ **File** ⇒ **Open** you could simply click on the second button in on the first (Standard) toolbar which looks like a yellow folder with an arrow pushing it open.

The left half of both the standard and formatting toolbars are common to the whole Microsoft Office suite of programs. The right halves of each have commands special to Excel. Buttons on the standard toolbar which are common to all programs are the **Copy**, **Paste** and **Undo** buttons. We will demonstrate the **Copy** operation.

Type the number 10 into cell A1 and press **Return**.

Now we will copy the number 10 and paste it to another cell.

Select cell A1 again, then click the copy button on the standard toolbar (denoted by two sheets of paper).

Now select any other cell on the worksheet. Click on the paste button (next to the copy button) and another number 10 should appear in your newly selected cell.

[7] Although the toolbars normally appear at the top of the screen they can actually be dragged around if this is more convenient. To do this, click the mouse pointer on the left ribbed area of the toolbar (not on a button), hold down and drag anywhere. It can be returned to its original position in a similar manner.

An invaluable button is the Undo button. Clicking on this undoes your last action or command. The small arrow on the right enables you to undo more than one previous action, in reverse order. Use with care, otherwise you may lose work you intended to keep. However the next arrow is the Redo button which will replace your work provided that you click it immediately.

Examples of buttons common to all programmes in the Office suite on the formatting toolbar are Bold, Italic and Underline. By selecting your original number 10 in cell A1 you can change its appearance by clicking on one, two or all of these three buttons in the formatting toolbar. When you click on these buttons they appear to remain depressed. By clicking on them again you can remove the formatting from your number.

2.11 Keyboard commands

Keyboard commands tend to be less intuitive because they have to be remembered. You can perform many commands without using the menu bar or toolbar, by holding down Control and pressing a letter or other character on the keyboard. Some of the keyboard command equivalents are displayed next to the relevant command on the menus from the menu bar, for example to perform a Save operation you can either do Menu ⇒ File ⇒ Save or press the keyboard Control–s.[8]

Other useful keyboard commands are:

Print	Control–p
Undo last operation	Control–z
Repeat last operation. This will be applied to whatever is selected.	Control–y

2.12 The formula bar

Next to the name box directly above the worksheet is the formula bar which has three sections :

[8] On the Macintosh use the Command key.

The right-hand section shows the current contents of the selected cell (in this case, the number 3035). If the cell contains a formula this will appear in the formula bar while the result will appear in the cell. This allows you to see the result and the formula behind it at the same time.

The centre section (grey) indicates whether the selected cell is being edited. Clicking in the right-hand section enables you to edit the contents, and while editing this centre section shows a red cross and green tick. Red obviously indicates "Cancel the edit" and green indicates OK.

The left-hand section shows the address of the currently selected cell (in this case A1). The down-arrow attached to this section will show any cell names in the workbook if you click it – we shall also discuss these later.

2.13 The status bar

The status bar appears at the bottom of the screen and indicates whether we are in Ready or Edit mode, and various other items including whether scrolling is locked on (this prevents the cell selection being moved with the arrow keys and is usually left off). Scroll lock is controlled by one of the labelled function keys, usually at the top right of your keyboard.

2.14 Saving workbooks

One of the most important matters to remember when using Excel (or any other program) is to save your work regularly. Computers crash for all sorts of reasons and while you may work for hours on many occasions without mishap one day you will lose a large amount of work because of a crash. Everyone has done it at some time and will do it again, but that is no reason to fail to remind you. Excel has an **Autosave** facility and it is recommended that this is turned on with **Menu** $\Rightarrow$ **Tools** $\Rightarrow$ **Autosave**. (If this command does not appear under the Tools menu you will need to install it as an add-in from your original program disk.)

If your workbook is new you will be prompted for a name. Enter a name in the highlighted space in the dialogue box. Excel will automatically attach information in order that the computer will recognise it as an Excel file in future.

You can save a second copy by using the **Menu** $\Rightarrow$ **File** $\Rightarrow$ **Save As** command. This will give you the same dialogue box – enter another name and this will become the new name of the workbook. The previous saved file will remain and can be opened again if desired. In this way you can save new versions of you work from time to time and revert to an earlier version if something goes badly wrong.

2.15 Help!

Excel has an interactive Help facility which can be very useful if you are not sure where to go next. To use this use the menu command **Menu** ⇒ **Help** ⇒ **Microsoft Excel Help**. An animated dialogue box will appear in which you can type a word or a question. Alternatively you can select **Menu** ⇒ **Help** ⇒ **Contents** and **Index** which will give a more conventional approach to finding information about Excel.

Chapter 3

Formulae and operators

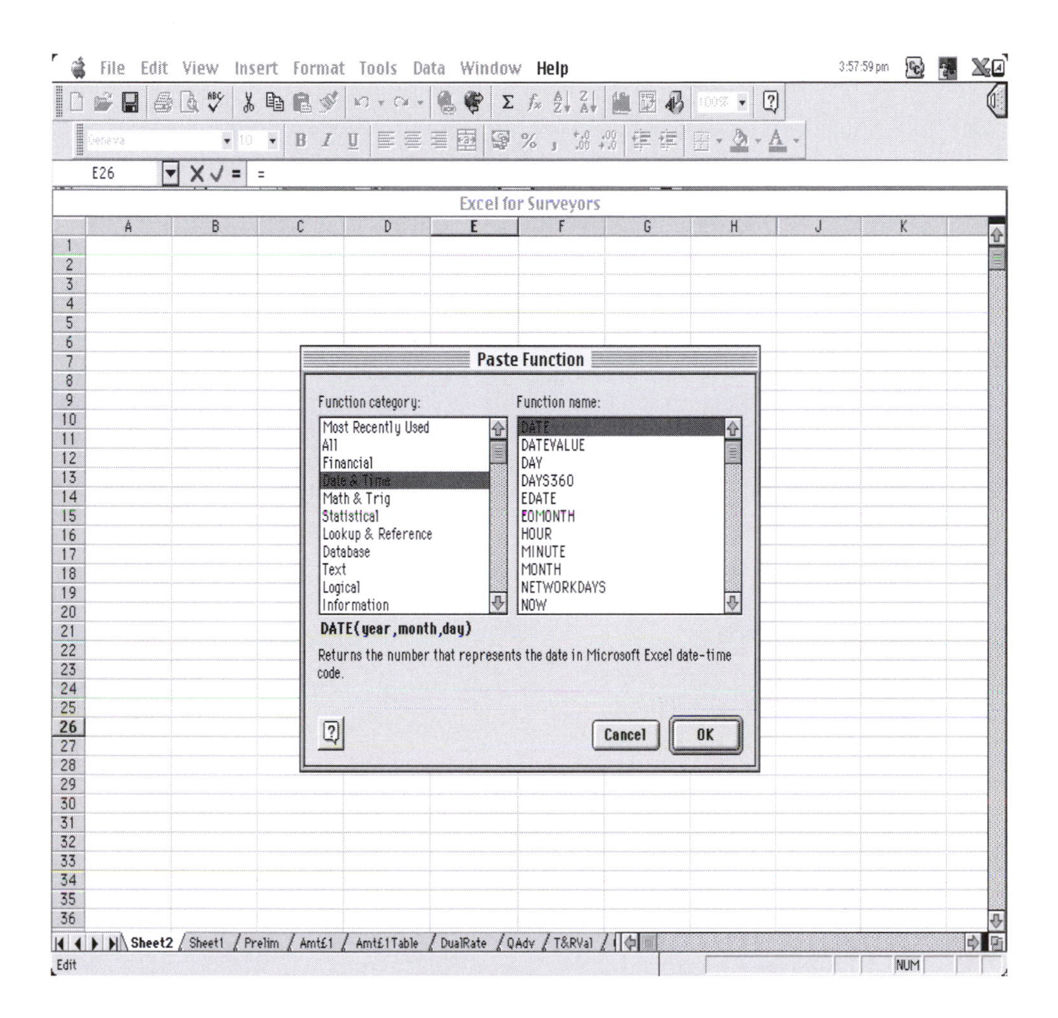

3.1 Formulae

Formulae are instructions to Excel to operate on the numbers that you have already entered. We inform the computer that it is an instruction instead of a number or text by putting an equals (=) sign in front of the instruction. Many instructions are in the form of equations.

3.2 Operators

Formulae are constructed using operators, and these will be used extensively. The principal operators are:

addition	+	multiplication	*
subtraction	–	division	/
exponentiation (raising to a power)	^	Boolean	True/False

Operators take precedence according to the normal conventions of mathematics – Boolean first, exponentiation next, then multiplication and division, and finally addition and subtraction. (We shall not consider Boolean operators here.) Parentheses "( )" takes precedence over all operators.

3.3 Writing formulae

Formulae in Excel generally use cell references, but may include numbers as constants. Example:

(a) Type 1.1 in cell A1 and 20 in cell A2
(b) Select cell A3 and type: = A1^A2 ("^" is above the 6 on the keyboard)[9]
Remember to press **Return**. The figure in cell A3 should now be 6.7275.
We can now use cell references instead of numbers to allow "what if" analysis.
(c) Select cell A2 and change the number to 30
The answer in cell A3 will immediately change to 17.4494.

3.4 Relative and absolute cell referencing

3.4.1 Relative cell referencing

Relative cell referencing allows you to apply the same formula to different cells in the worksheet without rewriting it.

Example

(a) Enter number 20 in cell A1, 20 in A2, 5 in B1 and 25 in B2.
(b) In A3 enter the formula

= A1 + A2

Remember to press **Return**.
(c) Select A3 again.
Now we will apply the formula in this cell to the next column of numbers by copying it across.
On the selection border around cell A3 is a small square in the bottom right corner of the cell. This is called the fill handle. Hover the mouse pointer on this, and the white selection cross will turn into a thin black cross. This is your copying cross.
(d) When this appears click and drag to cell B3 and let go.

Click on cell B3 and look at the formula behind it in the formula bar. The formula stays the same but the references it uses are different. The computer does this by default if you copy a formula. It is called relative cell referencing because it moves

---

[9] Hold down the SHIFT key to obtain characters above the numerals on the keyboard.

all the references in your formula by the same direction and distance as the formula itself.

3.4.2 *Absolute cell referencing*

You can stop it doing this. Using the previous example we will stop it changing A1 next time we copy.

(a) First select cell B3 and press the **Delete** key to clear it.
(b) Select cell A3, and click in the formula bar (at the top of the screen) so that we can edit the formula.
(c) Move the cursor so that it is flashing between the A and the 1.
(d) Press F4, at the top of your keyboard.[10]
This puts dollar signs against the reference A1, and fixes it in the formula so that it does not change when the formula is copied.
(e) Press **Return** and then copy the formula across as in step (d) of the previous example.

The formula in cell B3 will now add cells A1 and B2.

3.5 Formulas view

We have noted that when a formula is entered in a cell and the Return key is pressed, the formula appears in the formula bar and the result in the cell. However it is frequently convenient to display the formulae in their cells as well, and this can be done with the **Menu** ⇒ **Tools** ⇒ **Preferences**[11] command. In the Preferences dialogue box which appears under the Window Options heading, click on **Formulas** and **OK**, and all formulae will appear in their respective cells. Note that, as formulae are frequently quite long, all column widths are doubled, although they can be adjusted in the usual way. To turn off formulas view, repeat the process and the worksheet will return to normal, although if you have adjusted column widths in formula view you will have to readjust them back again.

(We have used formulas view in setting up several of our examples but it is not essential for you to always do this.)

[10] On the Macintosh press Command–t.

[11] Options, instead of preferences, on the PC.

Chapter 4

Formatting cells

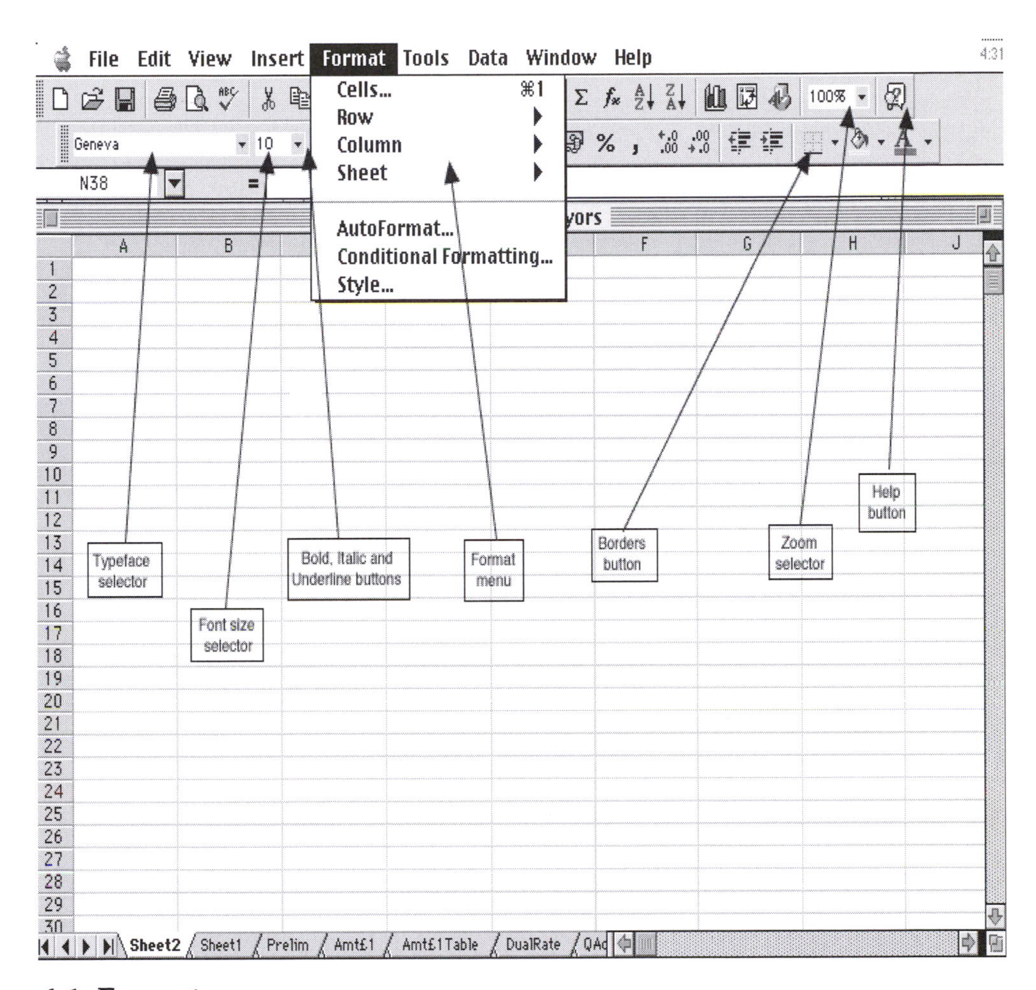

4.1 Formats

Formatting is the term used for changing the appearance of text or numbers to assist in making data more meaningful. We introduced formatting when we discussed the toolbars. For example we could use bold or a larger font for totals; put borders around tables; give numbers currency notation; or make decimal numbers appear as percentages. The Shift key and the Caps Lock key operate in the same way as on a typewriter, and enable capitals and other characters to be typed. It is possible to apply the same formatting to several non-adjacent cells at a time by holding down the Control key while selecting them.[12]

---

[12] On the Macintosh, the Command key.

Although Excel is mainly about numbers we will also need to add text to our work in order to explain it. Text in Excel is typed just as you would on a word processor, but the main purpose of the text is to label data.

Often your text will not fit into one cell, but there are several possible ways of getting around this problem.

If there is no entry in the next column the text will overflow into it. However if anything is subsequently entered into the next column the text will be truncated. Increase the column width either by dragging the division markers in the column header or by using the **Menu** ⇒ **Format** ⇒ **Column** ⇒ **Width** command and entering the width desired by number of characters or digits.

Use the **Menu** ⇒ **Format** ⇒ **Cells** ⇒ **Alignment** command and click the Wrap Text box. The column height will increase to allow for the additional text.

4.2 Emphasis – typeface and font size

We have shown you how to add emphasis to your data using the bold, italic, and underline commands on the toolbar (see Chapter 2.10). You can also change the typeface (the style of the characters) or the font size. On the left of the formatting toolbar is a box with the current typeface, and if you click on the adjoining arrow you will see a drop-down list of all the available typefaces. Similarly, there is an adjoining box and arrow showing the current font size and an arrow to permit you to see a selection of font sizes.

4.3 Justification

You can change the position of the text or number within the cell with the justification buttons. By default text is left-justified and numbers are right-justified.

4.4 Number formatting

On the formatting toolbar there is a currency button with a picture of money on it. Select one of your cells with a number in and apply this format by clicking on this button. A pound sign will precede the number and two decimal places will be added for the pence.

To change this format to accounting style, click the **Comma-Style** button.

To change the number of decimal places click on either the increase or decrease decimal button.

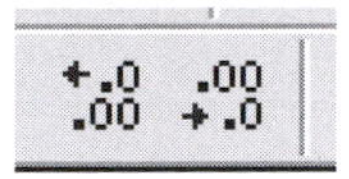

Percentages take a little more consideration. If you manually type the percentage sign (above the 5 on your keyboard) after a number the computer will recognise your number as a fraction of 100 and mathematically will interpret 5% as .05. However if you select a cell with a number already in it and click the **Percentage Button** (next to the currency button), it will add the percentage sign for you but will also expresses that number as a percentage. Thus if the cell already contains "5" it will be converted to 500%.

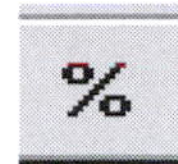

4.5 Cell formatting

Finally you might want to do a little formatting to the cell itself. Three useful buttons for this are the border, the fill colour and the text colour. These have drop down arrows next to them similar to the font and font size list boxes. By clicking on these you can choose whether you want to put a border all the way around the cell, or just one side; or change the background colour of your cell; or the colour of the text. A particularly useful facility here is to format a border at the bottom of a cell to indicate the total of a column of numbers.

You can also set borders with the **Menu** ⇒ **Format** ⇒ **Cells** ⇒ **Border** command which gives many more options in the style of formatting, including broken lines and variable thicknesses.

You may like to experiment with these.

4.6 Date and time

These can be useful to record, for example, the dates of transactions where these are being entered into a database. They have the advantage that, assuming the computer clock is set correctly, the day name, day and month in the current year or any combination can be entered with just two numbers separated by a slash. For example :

Select a cell and then **Menu** ⇒ **Format** ⇒ **Cells** ⇒ **Number** ⇒ **Date**. This displays the Format Cells dialogue box which allows you to set the format you wish, including the ability to define your own.

In the category box select Date, and then select a format from the **Type** column, for example scroll down and select "4-Mar-97" and **Return**.

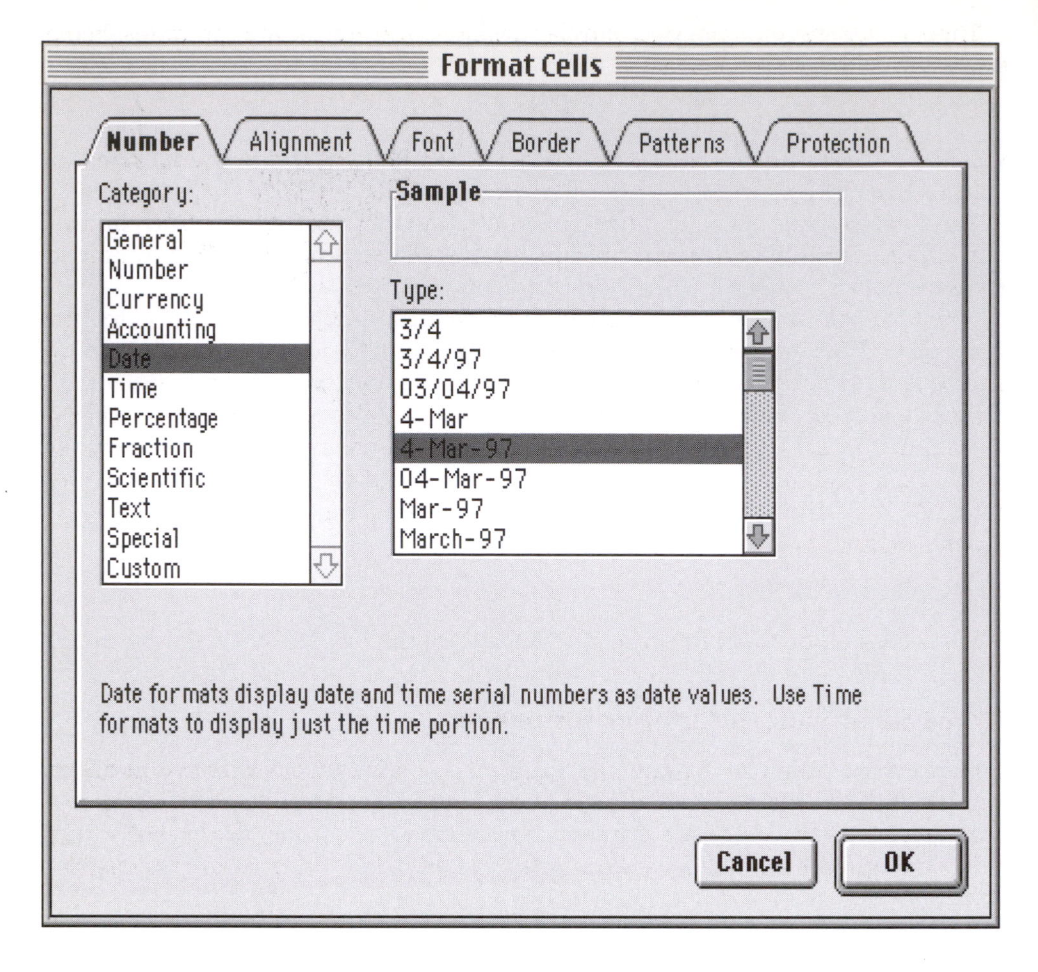

Now enter 8/9 in the cell. The result will be "8-Sep-00". The current year is the default, (although you can override this with, say, 8/1/99 for an earlier year.)

You may prefer your own format for the date. In this case select Category **Custom**. In the top line of the Type box enter

d mmmm yyyy

d	8	m	1	y	00
dd	08	mm	01	yy	00
ddd	Fri	mmm	Jan	yyy	2000
dddd	Friday	mmmm	January	yyyy	2000

and **Return**. The result will now be 8 January 1999.

More generally, date formats are interpreted as follows:

For example, an abbreviated date can be shown by the code

d mmm yy

which gives 8 Sep 99, and you can make up your own formats in similar fashion.

Time functions operate in a similar fashion, the hours and minutes being separated with a colon. Format

hh:mm:ss

will show hours minutes and seconds. Both date and time functions can appear in one cell, though this would need rather wide columns.

You can also enter the date function =NOW( ) and the current date and/or time will appear, depending on the formatting of the cell (the brackets in this command are essential).

If you format a cell as d mmm yy and another as hh:mm:ss and enter the function =NOW() into each , and then print the sheet, it will show the precise date and time of printing, as it will be updated on each occasion automatically.

Finally, you can do calculations based on dates since the computer converts the dates into numbers, the integer part representing the day number since 1 January 1900[13] and the fractional part the time. For example, format A1 and A2 as dates and enter 1/1 into cell A1 and 2/5 into A2 (for the dates 1 January and 2 May). Format A3 as a number and enter

=A2–A1.

The answer will be the number of days from 1 January to 2 May.

4.7 More advanced number formatting

If you use the **Menu ⇒ Format ⇒ Cells ⇒ Number** command several dozen formats will be offered, including for example an option to show negative numbers in red (very useful for financial calculations). You can also specify your own coding, and we shall use examples of this again later when we discuss valuations.

To specify your own code, the rules are:

(a) Text, including spaces, must be enclosed in inverted commas
(b) Numbers are represented by hash marks (###)
(c) A decimal point may be included within the hash marks to indicate the number of decimal places. For example, #,###.## would indicate commas for thousands separators and two decimal places.
(d) A zero will act as a placeholder. For example 0.## will show a zero before the decimal point in the absence of a whole number; 0.000 will show a zero and three places of decimals; 0.### will show a zero and up to three places of decimals if they exist.

The following could be used to label the number of years in a Present Value calculation:

"Present Value of £1 in " ## " years"

---

[13] By default the PC adopts 1 January 1900 as its starting date, whereas the Macintosh defaults to 1 January 1904. Either system can be converted to the other with **Menu ⇒ Tools ⇒ Options ⇒ Calculation**. We recommend that the default is not changed.

This will be interpreted as a number by the computer, and can be used in a calculation just as any other number. Note however that it will not overflow into the next column, nor can it be word-wrapped.

4.8 Comments

It frequently happens that we wish to make a note about the contents of a particular cell. This can be done very conveniently by selecting the cell and using the **Menu ⇒ Insert ⇒ Comment** function. A yellow "post-it" box appears alongside the cell into which any comment can be typed. Then click on any other cell and the comment box will disappear, leaving a red indicator in the corner of the cell. To read the contents of the note, place the cursor over the cell without clicking and the note will appear again.

To amend the note, select the cell and then **Menu ⇒ Insert ⇒ Edit Comment**. To remove the note select the cell and **Menu ⇒ Edit ⇒ Clear Comment**.

If you wish to see all notes together select **Menu ⇒ Tools ⇒ Options ⇒ View** and then **Comment & Indicator**. All comments on the worksheet will be shown.

If you use comments very frequently you can open the Reviewing Toolbar with **Menu ⇒ Insert ⇒ Toolbars ⇒ Reviewing** which gives useful options for viewing all comments.

Chapter 5

Functions

5.1 Functions

In mathematics, a function may be thought of as a black box into which a number or numbers are put, and out of which comes another number.[14] Inside the box are the instructions on what to do with the numbers – the formula. For example, AVERAGE is a function, and the formula concerned is:

$$\frac{\text{Sum}}{\text{Number of items}}$$

Excel contains a set of standard functions which can be found in **Menu ⇒ Insert ⇒ Function**. However you will frequently want to write your own formulae to carry out functions, and we will show you how to construct more examples later.

When you select standard functions a dialog box will appear, asking you to choose a function. There are two lists: **Category of Function** on the left, and **Name of Function** on the right. By clicking on a category in the left list, the relevant functions under that heading will appear on the right. Click on a function on the right and a description and the syntax for that function will appear at the bottom of the dialog box. Each function will begin with an "=" sign. The operations are hidden inside the function name, and a request for a range of cells, called the argument, will appear in brackets at the end. For example, if you select AVERAGE, Excel will respond with

AVERAGE(number1,number2,...)

AVERAGE is the function (usually referred to elsewhere as the "mean") and Excel is asking for the range of numbers to be inserted between the brackets. However you are helped along with this by another dialogue box which allows you to drag across the cells that contain the numbers you want to average.

The most relevant function categories for us as surveyors are Financial, Date & Time, and Statistical.

5.2 Financial functions – a note

There are nearly 100 financial functions, but most of them are of little interest to surveyors in this country. They are designed primarily for the American investment market where methods, characteristics and terminology are different from the British system. We shall discuss a few specific functions later but apart from these we recommend that you use the formulae which we will explain.

---

[14] The computer itself may be though of as a function black box – we enter information with the keyboard and something comes out on the screen or on the printer. Millions of operations happen inside the box though we do not need to know anything about them.

5.3 Statistical Functions[15]

Statistics is concerned with finding patterns and characteristics of sets of data, for example trends in office rents over the last 10 years. A full discussion is beyond the scope of this volume, but we give the following by way of example.

The values in a data set are referred to as a range. When Excel asks you to select a range it will mean the lowest cell reference to the highest cell reference in which your data set has been input, in whatever order that happens to be.[16]

Example

Find the mean (average), maximum, minimum and median of the numbers in Table 1.

The maximum, minimum median and average values of a set of numbers can be found using the AVERAGE MAX, MIN, and MEDIAN functions. All follow a similar syntax. For the mean of the numbers in your range the syntax will appear as:

= AVERAGE (number1,number2 . . .)

(a) Enter the numbers from Table 1 into column A of the worksheet.
(b) Enter the names of the functions into Column B (for reference).
(c) Enter the functions into Column C by selecting **Menu** ⇒ **Insert** ⇒ **Function** ⇒ **Statistical**. As you select each function you will be asked for the range – drag over the range to insert it in the upper box (ignore the lower box for this example).

The worksheet entries should appear as follows in formula view :

	A	B	C
1	Table 1		
2	865		
3	96	Average	=AVERAGE(A2:A12)
4	479	Maximum	=MAX(A2:A12)
5	597	Minimum	=MIN(A2:A12)
6	994	Median	=MEDIAN(A2:A12)
7	89		
8	673		
9	60		
10	916		
11	726		
12	497		

[15] There are dozens of books on statistics from the introductory and barely informative to the in-depth and barely comprehensible. A good general summary may be found in *Encyclopaedia Britannica* under "Statistics".

[16] When a dialogue box appears it may hide your numbers. You can move the box by clicking on its grey area, holding down the mouse button, and dragging.

The computed results in normal view are:

	A	B	C
1	Table 1		
2	865		
3	96	Average	544.727273
4	479	Maximum	994
5	597	Minimum	60
6	994	Median	597
7	89		
8	673		
9	60		
10	916		
11	726		
12	497		

The above functions really only exhibit superiority over the calculator when large sets of numbers are used (here you could easily see which of the numbers is the maximum). However among these functions are a host of valuable calculations that the computer is able to perform that make it invaluable. Again it is unlikely you will ever use all of them. The statistical category is after all the biggest, but once you are familiar with a few functions it is easy to get to grips with others as and when you need them.

Chapter 6

Valuation tables

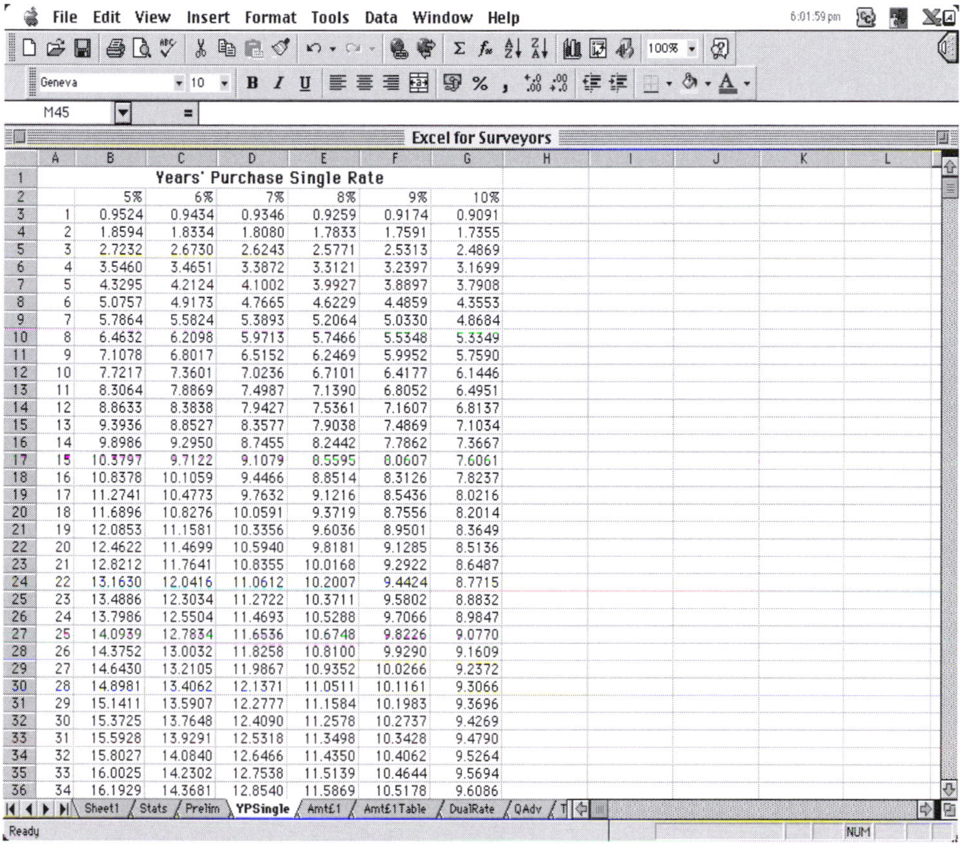

Excel for Surveyors

Years' Purchase Single Rate

	5%	6%	7%	8%	9%	10%
1	0.9524	0.9434	0.9346	0.9259	0.9174	0.9091
2	1.8594	1.8334	1.8080	1.7833	1.7591	1.7355
3	2.7232	2.6730	2.6243	2.5771	2.5313	2.4869
4	3.5460	3.4651	3.3872	3.3121	3.2397	3.1699
5	4.3295	4.2124	4.1002	3.9927	3.8897	3.7908
6	5.0757	4.9173	4.7665	4.6229	4.4859	4.3553
7	5.7864	5.5824	5.3893	5.2064	5.0330	4.8684
8	6.4632	6.2098	5.9713	5.7466	5.5348	5.3349
9	7.1078	6.8017	6.5152	6.2469	5.9952	5.7590
10	7.7217	7.3601	7.0236	6.7101	6.4177	6.1446
11	8.3064	7.8869	7.4987	7.1390	6.8052	6.4951
12	8.8633	8.3838	7.9427	7.5361	7.1607	6.8137
13	9.3936	8.8527	8.3577	7.9038	7.4869	7.1034
14	9.8986	9.2950	8.7455	8.2442	7.7862	7.3667
15	10.3797	9.7122	9.1079	8.5595	8.0607	7.6061
16	10.8378	10.1059	9.4466	8.8514	8.3126	7.8237
17	11.2741	10.4773	9.7632	9.1216	8.5436	8.0216
18	11.6896	10.8276	10.0591	9.3719	8.7556	8.2014
19	12.0853	11.1581	10.3356	9.6036	8.9501	8.3649
20	12.4622	11.4699	10.5940	9.8181	9.1285	8.5136
21	12.8212	11.7641	10.8355	10.0168	9.2922	8.6487
22	13.1630	12.0416	11.0612	10.2007	9.4424	8.7715
23	13.4886	12.3034	11.2722	10.3711	9.5802	8.8832
24	13.7986	12.5504	11.4693	10.5288	9.7066	8.9847
25	14.0939	12.7834	11.6536	10.6748	9.8226	9.0770
26	14.3752	13.0032	11.8258	10.8100	9.9290	9.1609
27	14.6430	13.2105	11.9867	10.9352	10.0266	9.2372
28	14.8981	13.4062	12.1371	11.0511	10.1161	9.3066
29	15.1411	13.5907	12.2777	11.1584	10.1983	9.3696
30	15.3725	13.7648	12.4090	11.2578	10.2737	9.4269
31	15.5928	13.9291	12.5318	11.3498	10.3428	9.4790
32	15.8027	14.0840	12.6466	11.4350	10.4062	9.5264
33	16.0025	14.2302	12.7538	11.5139	10.4644	9.5694
34	16.1929	14.3681	12.8540	11.5869	10.5178	9.6086

Sheet1 / Stats / Prelim / YPSingle / Amt£1 / Amt£1Table / DualRate / QAdv

6.1 Creating valuation tables

The first valuation tables were constructed using paper and pencil, with a lot of hand-written calculations. Later, the best-known tables in the property world, *Parry's Valuation Tables*, were calculated by means of a mechanical adding and subtracting machine. Multiplication and raising to powers was a matter of repetitive addition and must have taken many hours and a good deal of physical effort.

Excel can produce a page of tables in less than a second, but the setup needs some thought in order to be efficient. The examples which follow demonstrate the use of names, formatting, and general layout considerations. Each example will produce a page similar to that of *Parry's*.

6.2 The amount of £1

This is the name given to the way in which an investment of £1.00 will grow if the annual interest is reinvested. The formula for this is commonly written as $(1+i)^n$ where i is the annual rate of interest and n is the number of years. The superscript n indicates "raise to the power of" and is often referred to as the "exponentiation operator".

In Excel (and other computer programs) this operator is indicated by a caret (^) instead of the usual superscript. Thus:

Textbook style	$(1.05)^7$
Worksheet style	=(1.05)^7

A single calculation of the Amount of £1 can readily be done on any pocket calculator which has a power function. The computer has the advantage that it will allow us to copy the formula and then to change the inputs (i and n) at will, eventually calculating a complete table of values similar to those found in *Parry*.

We can calculate the result for single inputs in one cell of the worksheet. Note the "=" sign. If this is not the first character of the formula Excel will consider it to be text and will not perform any calculation (try it yourself and see). To find the amount of £1 for 7 years at 5% we type into any cell

=1.05^7

Before we construct a complete table we shall extend the previous example by entering the number of years and the rate of interest into separate cells and then using a third cell to complete the calculation. Thus we enter

(a) 5% into cell B2. (Remember that typing "5" and "%" will automatically format the cell as percentage and internally the number is stored as 0.05.)

(b) 7 into cell C2

(c) the formula into cell D2. This formula follows the normal style, modified to refer to the other two cells.

The computer will see these entries like this:

	A	B	C	D
1				
2		0.05	7	=(1+B2)^C2
3				

The final result will appear on the screen as:

	A	B	C	D
1				
2		5%	7	1.407100423
3				

With the data now entered we have the same answer in cell D4 as previously, but now we are able to change the rate of interest or the number of years independently and obtain any value we wish. Of course we can add text to indicate what each of the numbers refers to.

6.3 Names

One very useful feature of Excel is the ability to name cells or groups of cells. Remember that each cell has its own reference, and this is the reference you use in formulae. Some of your formulae may become quite long; others may be brief but with lots of them. It may be necessary to examine the way the formula was constructed long after you have written it, or it might be difficult in large workbooks to find the reference of the number you want to include in your formula. Naming cells can solve this problem.

Example

(a) Select cell B2
(b) Click on **Menu** ⇒ **Insert** ⇒ **Name** ⇒ **Define**. In the top section of the Define Name box you enter the name "interest". At the bottom you will see the reference of the cell you have selected.
(c) Select cell C2. Similarly define the name as "years" and press return. We can now use these names in our formula.
(d) Select cell D2 and type the formula =(1+interest)^years.

You should have the Amount of £1 as before.

6.4 Constructing a table of valuation factors

Names are particularly useful in compiling a table of valuation factors. Suppose we wish to construct a table of the Amount of £1 with rates of interest from 5% to 10% and for years from 1 to 50. Using a new workbook, we can assign the name "interest" to the top row where the rates of interest will be entered, and the name "years" to the left-hand column where the years will be inserted. As before, to enter a name go to **Menu** ⇒ **Insert** ⇒ **Name** ⇒ **Define**. Enter "interest" in the top box. Click in the bottom box and drag over B2:G7 where you will enter rates of interest. Click OK and the name will be established. Name cells A3:A52 as "years" in the same way.

Nothing will as yet appear on the worksheet.

[CAUTION – These names will apply to the whole workbook so if you use other worksheets in the same workbook a formula may pick up a name from a different worksheet. This may not be what you intended. For this reason, in the examples which follow, we recommend that you do each example in a different workbook.]

You can check that you have the correct cells by using the **Menu** ⇒ **Edit** ⇒ **Goto** command. The names you have entered will be shown and if you click on a name the relevant cells will be selected.

We now enter the rates of interest along the top row, from 5% to 10% and the left-hand column we enter the years from 1 to 50. However we can use a short cut to enter the years instead of typing each number individually, as follows.

Enter 1 in cell A3 as the first year in the "years" column.

Go to **Menu** ⇒ **Edit** ⇒ **Fill** ⇒ **Series** and in the box which pops up select :

(a) the **Columns** button
(b) the **Linear** button
(c) step value 1
(d) stop value 50
(e) Press OK and presto – you have all the years at one go!

Now that we have interest rates across the top and years down the first column we can insert the formula which we had before in cell B3 :

=(1+interest)^years

The Amount of £1 for one year at 5% will appear.

To complete the table, keep this cell selected and use the **Menu** ⇒ **Edit** ⇒ **Copy** command. The cell border will now be flashing. Next, select the whole of the table by dragging from B3 to G52 and select **Menu** ⇒ **Edit** ⇒ **Paste**. The entire table will be filled with the Amount of £1 for each year and rate of interest. Another Presto!

The formula in the formula bar is now in each of the cells. Each cell is calculated by looking at row 2 for the appropriate rate of interest and column A for the number of years.

Finally, to tidy it up, we can fix all the results to the same number of decimal places and insert a suitable title. The following steps set decimal places :

(a) Again select the table from B3 to G52 (ie all the calculated factors).
(b) Select **Menu** ⇒ **Format** ⇒ **Cells**. The **Format Cells** dialogue box appears. Select the **Number** tab and then **Custom**. In the box under the word **type** enter "0.0000" (for four decimal places).
(c) Click OK and the entire table of values will be adjusted to four places. Next we create a title in Row 1.
(d) In cell A1 enter "The Amount of £1"
(e) As an optional extra, click on **Format** ⇒ **Cells** ⇒ **Font** and choose a typeface and size which you prefer.
(f) Select cells A1:G1 and click on the Merge and Centre button This will centre the title over the whole table.

Finally we can adjust the column widths by dragging the dividers in the column headers, or by selecting the columns and using the **Menu** ⇒ **Format** ⇒ **Column** ⇒ **Width** command to specify the width (remember that the number refers to the number of digits and not a measurement). The first few rows of the finished table will appear as shown opposite.

6.5 The Present Value of £1

The construction of this table is very similar to The Amount of £1. The formula is commonly written as or $(1+i)^{-n}$. However for Excel it is re-written as:

=(1+i)^–n

	A	B	C	D	E	F	G
1				**The Amount of £1**			
2		5%	6%	7%	8%	9%	10%
3	1	1.0500	1.0600	1.0700	1.0800	1.0900	1.1000
4	2	1.1025	1.1236	1.1449	1.1664	1.1881	1.2100
5	3	1.1576	1.1910	1.2250	1.2597	1.2950	1.3310
6	4	1.2155	1.2625	1.3108	1.3605	1.4116	1.4641
7	5	1.2763	1.3382	1.4026	1.4693	1.5386	1.6105
8	6	1.3401	1.4185	1.5007	1.5869	1.6771	1.7716
9	7	1.4071	1.5036	1.6058	1.7138	1.8280	1.9487
10	8	1.4775	1.5938	1.7182	1.8509	1.9926	2.1436
11	9	1.5513	1.6895	1.8385	1.9990	2.1719	2.3579
12	10	1.6289	1.7908	1.9672	2.1589	2.3674	2.5937
13	11	1.7103	1.8983	2.1049	2.3316	2.5804	2.8531
14	12	1.7959	2.0122	2.2522	2.5182	2.8127	3.1384
15	13	[illegible]	[illegible]	[illegible]	[illegible]	[illegible]	[illegible]

To construct this table copy the previous table to a new workbook and delete the Amount of £1 factors (cells B3 : G51) with **Menu ⇒ Edit ⇒ Clear ⇒ Contents**. The process is then very similar to the Amount of £1.

Enter the formula into cell B3 using the names as previously discussed as:

=(1+interest)^–years

Copy the formula and paste it to cells B3 : G52 as before. (Note the minus sign in the formula.)

Select cells B3 : G52 and set the number of decimal places. You may wish to use an extra place because the Present Value is always less than unity and falls to quite low values for higher rates of interest and numbers of years.

6.6 Years' Purchase, single rate

Single rate tables may also be constructed in a similar manner. Make a copy of the original Amount of £1 table in a new workbook and clear cells B3 : G52. Then insert the next formula and proceed as before. This formula is a little more complicated, being commonly written as:

$$\frac{1 - (1 + i)^{-n}}{i}$$

For the purposes of Excel we rewrite it as:

=(1–(1+interest)^–years)/interest[17]

[17] A Quarterly in Advance version of Years' Purchase may be constructed by modifying the denominator of the formula as follows:

$$\frac{1 - (1 + i)^{n}}{4(1 - (1 + i)^{-1/4})}$$

The Excel version of this is:

= (1–(1+interest)^–years)/(4*(1–(1+interest)^(–1/4)))

Note that great care must be taken to place the brackets in exactly the right position, otherwise there will be an error in the result. If they do not match correctly the formula will give the wrong answer or not calculate at all.

6.7 Years' Purchase of a reversion to a perpetuity

The formula is conventionally written as:

$$\frac{(1 + i)^{-n}}{i}$$

By analogy with the previous examples, the computer version is

=(1+interest)^–years/interest

Enter this in cell B2, copy to the body of the whole table as before, and we have another page of tables.

6.8 The Amount of £1 pa,[18] the Annuity £1 will purchase and the annual sinking fund

Following the same procedure, enter the relevant formulae for these functions, which are:

	Conventional form	Computer form
Amount of £1 pa	$\frac{(1 + i)^{n} - 1}{i}$	=((1+interest)^years–1)/interest
Annuity £1 will purchase	$\frac{i}{1 - (1 + i)^{-n}}$	interest/(1–(1+interest)^–years)
Annual sinking fund	$\frac{i}{(1 + i)^{n} - 1}$	=interest/((1+interest)^years–1)

6.9 Dual rate

This formula is more complicated as we have an additional variable – the sinking fund rate – and possibly a second additional variable – the income tax allowance. Since these are usually regarded as constants for the purposes of a valuation table, it is convenient to define them separately at the top of the sheet and to follow the previous convention of years in rows and remunerative rates across the columns.

---

[18] For very large values of n and i the result is likely to be too large to fit into the column, and it will be replaced with hash marks (#############). In this case you may need to reformat individual cells by reducing the number of decimal places.

We could of course define separate tables for YP dual rate with and without allowance for income tax, but a single table will suffice if we set the income tax rate to zero if this allowance is not required.

The normal equation for the Dual Rate YP allowing for income tax is:

$$\frac{1}{i + \frac{s}{(1 + s)^n - 1} * \frac{1}{1 - t}}$$

The sinking fund rate in C2 will be named "asf" and the tax rate in F2 "tax", and the formula translated into Excel style becomes:

=1/(interest+(asf/((1+asf)^years–1))*(1/(1–tax)))

To set up the table we can conveniently set the asf and tax rates at the top of the table in cells B2 and F2 respectively, and name them "asf" and "tax". The body of the table will be similar to those previously discussed, except that there will be one more row at the top of the table.

	A	B	C	D	E	F	G
1	YEARS' PURCHASE DUAL RATE						
2		Asf rate	3%		Tax rate	40%	
3		Remunerative rate					
4		5%	6%	7%	8%	9%	10%
5	1	0.5825	0.5792	0.5758	0.5725	0.5693	0.5660
6	2	1.1481	1.1351	1.1223	1.1099	1.0977	1.0858
7	3	1.6972	1.6688	1.6415	1.6149	1.5893	1.5644
8	4	2.2303	2.1816	2.1350	2.0904	2.0476	2.0065
9	5	2.7478	2.6743	2.6047	2.5386	2.4757	2.4159
10	6	3.2503	3.1480	3.0519	2.9615	2.8764	2.7959
11	7	3.7382	3.6035	3.4781	3.3612	3.2519	3.1495
12	8	4.2118	4.0416	3.8846	3.7393	3.6045	3.4791

6.10 The all-risks yield and implied growth

In recent years the concepts of the capitalisation rate or all-risks yield, and the equated yield or target rate, and the relation between these and anticipated future growth have received much attention. In addition the Investment Property Forum has recently made further proposals for the calculation of interest rates. While we do not consider the theory here, we will demonstrate the construction of simple tables of the all-risks yield and implied growth.

The relevant formulae are as follows

All-risks yield or capitalisation rate

$$k = r * \left[1 - \frac{(1 + g)^n - 1}{(1 + r)^n - 1}\right]$$

Implied future growth

$$g = \sqrt[n]{1 + (1 - \frac{k}{r}) * ((1 + r)^n - 1)} - 1$$

In the table of capitalisation rates below we allocate the following names :

C2	review
A5:A54	equated_yield
B4:G4	growth

The formula for the body of the table is then

=equated_yield*(1–((1+Growth)^review–1)/((1+equated_yield)^review–1))

The table of Implied Growth is constructed in a similar manner except that cells B4:G4 are named "all_risks". The formula for the body of the table is then :

=(1+(1–all_risks/equated_yield) *((1+equated_yield)^review–1))^(1/years)–1

[Caution - if you construct these tables in the same workbook use different names, such as "review2" for C2, and equated_yield2 for A5:A54, otherwise there will be a conflict.]

Excel for Surveyors

	A	B	C	D	E	F	G
1	IMPLIED GROWTH						
2		Review	5 years				
3		All-risks yield					
4	Equated yield %	4%	5%	6%	7%	8%	9%
5	9.50%	5.908	4.930	3.913	2.855	1.751	0.597
6	10.00%	6.441	5.472	4.467	3.421	2.331	1.192
7	10.50%	6.973	6.014	5.020	3.986	2.909	1.785
8	11.00%	7.505	6.556	5.571	4.549	3.485	2.376
9	11.50%	8.036	7.096	6.122	5.111	4.060	2.964
10	12.00%	8.566	7.635	6.672	5.672	4.633	3.550
11	12.50%	9.096	8.174	7.220	6.231	5.204	4.135

6.11 Calendar

Finally, although not strictly a valuation table, since many leases are for long periods of years, a simple but useful long-term calendar can be constructed very easily as follows:

(a) In a new worksheet, enter a title, if desired, in row 1
(b) Enter the following into cells A2:G4

Excel for Surveyors

	A	B	C	D	E	F	G
1	Calendar						
2	=A3	=A2+1	=B2+1	=C2+1	=D2+1	=E2+1	=F2+1
3		=A3+1	=B3+1	=C3+1	=D3+1	=E3+1	=F3+1
4	=A3+7	=B3+7	=C3+7	=D3+7	=E3+7	=F3+7	=G3+7
5							

Most of these cells can of course be copied from an initial entry.

(c) Format Row 2 as "dddd" and Rows 3 onwards as "d mmm yyyy".
(d) Now enter your start date into A3. We will enter 1/1/2000.

In normal view your sheet will now appear as:

Excel for Surveyors

	A	B	C	D	E	F	G
1				**Calendar**			
2	Saturday	Sunday	Monday	Tuesday	Wednesday	Thursday	Friday
3	1 Jan 2000	2 Jan 2000	3 Jan 2000	4 Jan 2000	5 Jan 2000	6 Jan 2000	7 Jan 2000
4	8 Jan 2000	9 Jan 2000	10 Jan 2000	11 Jan 2000	12 Jan 2000	13 Jan 2000	14 Jan 2000
5	[illegible]	[illegible]	[illegible]	[illegible]	[illegible]	[illegible]	[illegible]

We have now created a calendar for the first two weeks of the Millennium year. To extend the calendar, select cells A4:G4, and drag down the small square in the bottom right-hand corner to create a calendar as long as you wish starting on 1 January 2000. To start your calendar on any other date change the date in A3 to that date, and all dates, including the day of the week at the head, will change.

Finally, you may wish to create larger cells so that your calendar can also include diary entries. Select from Row 3 to the end of your calendar and click on one of the lines dividing the row numbers. The height of the row in points will appear immediately above, and you can drag down to enlarge the cells. A cell height of 50 points might be a reasonable size.

You can of course format the columns in many different ways, including applying a shading to the Saturday and Sunday columns. Or you could change to Landscape view and widen the columns. Before printing, make sure that the margin settings allow all the columns to appear on one sheet.

You can also create a backdated calendar, back to 1 January 1904 or 1 January 1900, depending on the date system on your computer.

Chapter 7

Standard valuations

7.1 Fully let properties

The most basic capitalisation of the income from a property let at full rental value on full repairing terms is usually as follows:

Rent received	8,000
YP perp, 8%	12.5
	100,000

This can easily be set up on the worksheet and in formula appears as follows:

	A	B	C
1	FULLY LET PROPERTY		
2	Rent		8000
3	YP Perp	0.08	=1/B3
4			=C2*C3

In normal view it appears as follows. Note that cell B3 is formatted as 0.#% so that one place of decimals will appear if entered.

	A	B	C
1	FULLY LET PROPERTY		
2	Rent		8,000
3	YP Perp	8%	12.5
4			100,000
5			

7.2 Term and reversion

This is still probably one of the most common styles of valuation, even though the theory has been the subject of many criticisms in recent years. These have mainly been concerned with the problems of implied growth and subjective estimation of the capitalisation rate, and this aspect will be considered later.

Example

Value the freehold interest in a shop which is let on lease with four years unexpired at a rent of £5,000 pa. The landlord is responsible for external repairs. The current rental value on the same terms is estimated to be £7,000 pa.[19]

[19] This valuation assumes the conventional practice of increasing the reversionary rate by 1%. For a critical analysis of this see Philip Bowcock *Journal of Valuation* Vol 1 No 4 pp 366–376.

The following is a typical term and reversion valuation.

Rent	£5,000	
less external repairs	500	
	4,500	
YP 4 years, 7%	3.3872	15,242
Reversion to	7,000	
less external repairs	500	
	6.500	
YP perp def 4 years, 8%	9.1879	59,721
		74,964

(There is an apparent error here due to rounding.)
Setting this up in Excel, formulas view appears as follows :

	A	B	C	D
1	TERM AND REVERSION			
2				
3	Rent		5000	
4	less external repairs		500	
5			=C3-C4	
6	4	0.07	=(1-(1+B6)^(-A6))/B6	=C5*C6
7	Reversion to		7000	
8	less external repairs		500	
9			=C7-C8	
10	4	0.08	=((1+B10)^(-A10))/B10	=C9*C10
11				=D6+D10
12				

Note that the formulae for YP and YP deferred are those we have already used to compile the tables. The two entries of "4" need to be formatted to indicate what they refer to and this is done with **Menu ⇒ Format ⇒ Number** as previously discussed.

After returning to normal mode the final result appears as:

	A	B	C	D
1	TERM AND REVERSION			
2				
3	Rent		5,000	
4	less external repairs		500	
5			4,500	
6	YP 4 years	7%	3.3872	15,242
7	Reversion to		7,000	
8	less external repairs		500	
9			6,500	
10	YP perp deferred 4 years	8%	9.1879	59,721
11				74,964
12				

We are now in a position to do "What if" modifications since each input item can be altered and the effect on the final answer can be seen immediately. Note that to change cell A6 and A10 we only enter the number of years, and the existing text formatting is retained. (See 4.7)

7.3 The layer method

This method assumes that the initial income continues in perpetuity and the increase on reversion is capitalised separately. In formula mode the worksheet appears as follows.

	A	B	C	D
1	**LAYER METHOD**			
2	Rent		5000	
3	less external repairs		500	
4			=C2-C3	
5	YP perpetuity	0.07	=1/B5	=C4*C5
6	Reversion to marginal incom		2000	
7	4	0.08	=((1+B7)^(-A7))/B7	=C6*C7
8				=D5+D7
9				

In normal mode, the result is:

	A	B	C	D
1	**LAYER METHOD**			
2	Rent		5,000	
3	less external repairs		500	
4			4,500	
5	YP perpetuity	7%	14.2857	64,286
6	Reversion to marginal income		2,000	
7	YP perp deferred 4 years	8%	9.1879	18,376
8				82,661
9				

Chapter 8

Some useful tools

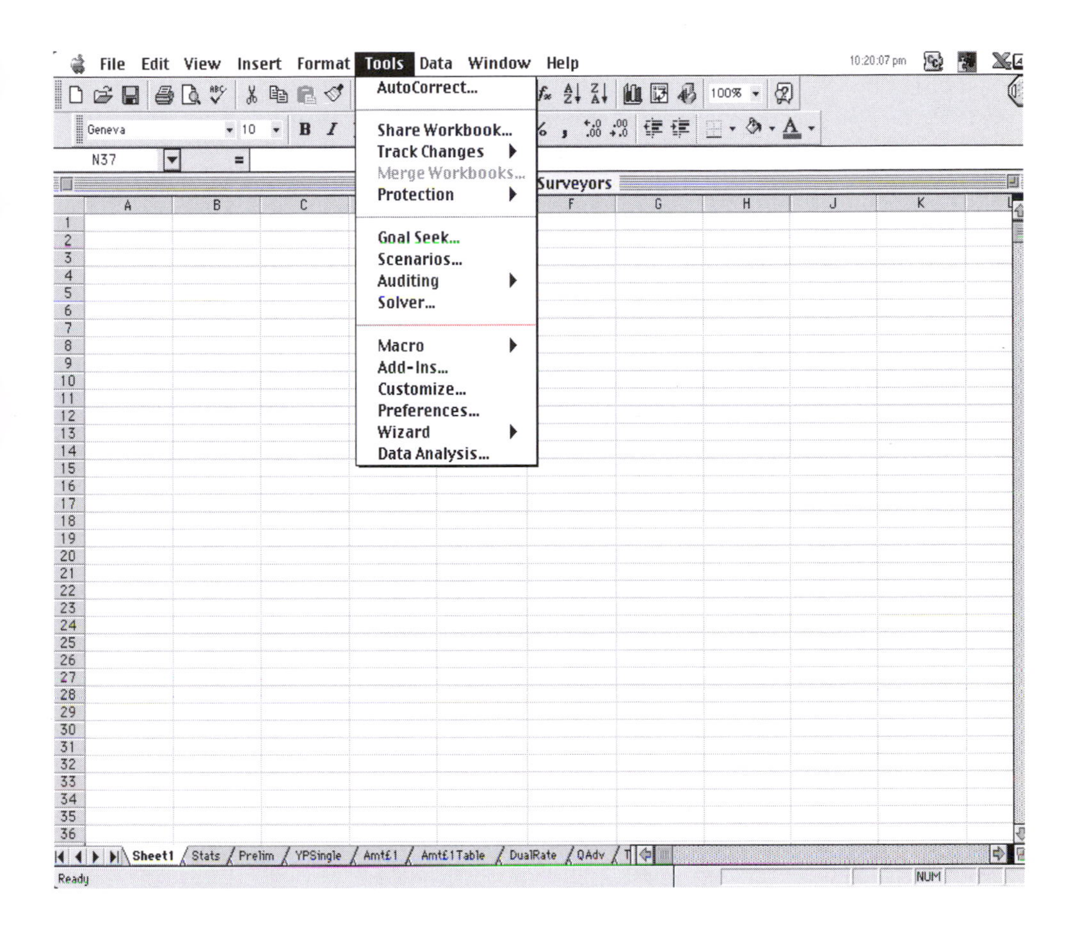

8.1 Spelling

This tool which is found in **Menu** ⇒ **Tools** is similar to the one used in Word (the word processor) and in fact uses the same dictionary. It will check all text in the spreadsheet except text which is part of a formatted number.

8.2 Auditing

One certainty about the use of worksheets is that you will never get it right on the first attempt, and you will be faced with one of the error messages, or a result which is clearly wrong. In some cases the problem is relatively easy to find and correct, such as where there is a missing parenthesis, but others may require tracing to another cell. Alternatively, we may be trying to understand the structure of a worksheet to discover what it is actually doing.

The **Auditing** buttons can be very useful here. To use them:

(a) Click on the cell which you want to investigate.

(b) Go to **Menu** ⇒ **Tools** ⇒ **Auditing** ⇒ **Show Auditing Toolbar**. A new toolbar will appear. If this is not in a convenient part of the screen you can drag it to a different position (see Chapter 2.10).

There are eight buttons on the toolbar. The first four allow you to trace and remove precedents (cells which contribute to the value in the selected cell) and dependants (cells which depend on the selected cell). The other four allow you to remove all arrows, to trace errors, and to attach and read information about the cell.

(c) Select a cell and click on the first and third buttons (Trace Precedents and Trace Dependants) to see the structure of the calculations.

Our previous example would show the following arrows with all formulae selected:

	A	B	C	D
1	TERM AND REVERSION			
2				
3	Rent		5,000	
4	less external repairs		500	
5			4,500	
6	YP 4 years	7%	3.3872	15,242
7	Reversion to		7,000	
8	less external repairs		500	
9			6,500	
10	YP perp deferred 4 years	8%	9.1879	59,721
11				74,964
12				
13				
14				
15				
16				

Auditing

8.3 Goal seek

How many times do you, as a valuer, think "I know the value of that but I need a calculation to back up my opinion". Excel has just the facility to help you do this with the "Goal Seek" function. This is found under **Menu** ⇒ **Tools** ⇒ **Goal Seek**.

Example

Suppose we are convinced that the capital value of the freehold of a shop is £500,000. The rent is £30,000 pa with five years unexpired, and the rental value if let now is £50,000. Using a traditional approach of term and reversion we want to find rates of interest which would produce this answer.

First we set up a valuation in standard format as before and guess the rates of interest which might be appropriate. In this case the entry in cell B6 is

=B3+.01

to ensure that the reversionary rate is 1% more than the term rate.

	A	B	C	D
1				
2	Income		30,000	
3	YP 5 years	7.00%	4.1002	123,006
4				
5	Reversion to		50,000	
6	YP perp deferred 5 years	8.00%	8.5073	425,364
7				548,370
8				

The rates selected are a good guess but not close enough, so we use **Goal Seek** to refine the calculation with the following steps:

(a) Select the cell (D7) which you want to be £50,000.

(b) Click on **Menu** ⇒ **Tools** ⇒ **Goal Seek** and you will see the Goal Seek dialogue box. Enter 500000 (no commas) in the "To value" box, click on the "By changing cell" box and then on Cell B3, and OK.

Goal Seek

Set cell:	D7
To value:	500000
By changing cell:	B3

Cancel OK

There will be some activity on the screen and the calculation will change to:

	A	B	C	D
1				
2	Income		30,000	
3	YP 5 years	7.69%	4.0254	120,762
4				
5	Reversion to		50,000	
6	YP perp deferred 5 years	8.69%	7.5848	379,238
7				500,000
8				

Obviously you can experiment with any of the other inputs (existing rent or reversionary rent) to achieve the desired result. This process can of course be adapted to any similar situation where you wish to adjust inputs to obtain the required answer.

Chapter 9

Discounted cash flow

9.1 Introduction

The difference between discounted cash flow (DCF) and traditional valuation methods is that in the former assumptions of rental growth and the required rate of return are explicit whereas in the traditional approach assumptions about growth are implied in the all-risks yield. A debate has continued since the 1970s property crash over whether traditional valuation techniques continue to be relevant.

The advantage of the cash flow approach is that:

(i) We can specify the target rate and therefore are able to make comparisons with the returns on other types of investments
(ii) We can specify the growth rate, and can make decisions as to whether we consider this will continue in the future. (Growth rates in the future cannot of course be predicted precisely.)

Despite facilitating a more analytical approach, DCF has probably not been favoured because of the greater number of components to the valuation.

The setting out of the cash flow details is critical. It is bad practice not to state your assumptions, and the information you are using before you set out the cash flow. Conveniently though, stating this information allows you to refer to it repeatedly when it is used more than once.

9.2 Setting up a discounted cash flow calculation

Example

Value an investor's interest in a freehold property which he purchased on 1 July for £1,250,000. He will receive a rent of £100,000 pa annually in arrears with five-year reviews. He intends to hold the property for five years. He seeks an overall return of 10% pa and rental growth is expected at about 3% pa.[20]

9.2.1 Setup

The initial setup is as follows.

Enter headings in row 6 and labels in cells B1:B4 as shown. Headings in row 6 are formatted as **Menu ⇒ Format ⇒ Cells ⇒ Alignment ⇒ Wrap Text** so that text is not cut off at the cell boundary.

Name Cells A7:A12 as "Year" and enter year numbers (0–5). Name Cell C4 as "NPV rate".

---

[20] For a consideration of the estimation of the market's expectation of growth, see the section 9.3 on over-rented property.

Enter formulae as follows:

Insert in	Formula	and
E7	=SUM(B7:D7)	Copy down to E12
B7	–C1	(Negative because it is a cash outflow)
C8	=C2	Copy down to C12
D12	=C3	
F7	=(1+NPVrate)^-year	Copy down to F12
G7	=E7*F7	Copy down to G12

This completes the initial setup. The worksheet will then appear as below.

Excel for Surveyors

	A	B	C	D	E	F	G
1		Purchase price					
2		Rent					
3		Sale price					
4		NPV Rate					
5							
6	Year	Purchase price	Rent	Sale price	Net Cash flow	PV factor	PV cash flow
7	0	0			0	1.0000	0
8	1				0	1.0000	0
9	2				0	1.0000	0
10	3				0	1.0000	0
11	4				0	1.0000	0
12	5			0	0	1.0000	0
13							0
21							

Assuming that the all-risks yield is unchanged at the end of the period the expectation of the sale price may be calculated as follows:

Rent at time of purchase	100,000	
Amount of £1, 5 years, 3%	1.1593	
Rental value at time of sale	115,927	
YP perp, 8%	12.5	
Estimated sale price	1,449,093	Say £1,500,000

We now enter the data:[21]

Cell	Subject	Entry
D1	initial purchase price	1250000
D2	rent	10000
D3	the estimated sale price	1500000
D4	the discount rate	10%

[21] It is not advisable to put static figures in your cash flow – always relate them to other data by way of formulae, however simple. If you change any of your inputs static data will not automatically change and you may get an error.

Excel for Surveyors

	A	B	C	D	E	F	G
1		Purchase price		1,250,000			
2		Rent		100,000			
3		Sale price		1,500,000			
4		NPV Rate		10.000%			
5							
6	Year	Purchase price	Rent	Sale price	Net Cash flow	PV factor	PV cash flow
7	0	-1,250,000			-1,250,000	1.0000	-1,250,000
8	1		100,000		100,000	0.9091	90,909
9	2		100,000		100,000	0.8264	82,645
10	3		100,000		100,000	0.7513	75,131
11	4		100,000		100,000	0.6830	68,301
12	5		100,000	1,500,000	1,600,000	0.6209	993,474
13							60,461
21							

Column G will now show the present value each of the cash flows expected during the next five years. Clearly this process can be continued for any number of later years.

9.2.2 Net present value

The NPV or investment worth is your valuation of the income from the property to this particular purchaser and is calculated by adding the NPV of the future income stream to the purchase price. This total of the discounted cash flows, can be calculated by adding up the present values of the cash in each year by selecting cell G13 and clicking the **Autosum** button. By default this will sum all the numbers immediately above. The answer, as shown above, should come to £60,461.

The NPV represents the algebraic sum of all the future cash flows at 10%. It follows that if the investor required a 10% pa overall return on the above investment he could have afforded to pay £60,461 above the asking price.

9.2.3 Internal rate of return

This is the overall annual return on a series of cash flows, that is, the discount rate which will give a Net Present Value of zero. In mathematical terms this calculation is impossible to solve directly, and the solution must therefore be found by a numerical approach of successive approximation ("trial and error").

Select a cell where you wish the answer to appear, select **Menu** ⇒ **Insert** ⇒ **Function**. Select the **financial** category and **IRR**. The dialogue box which appears will ask for the cash flow and a guess of the solution. Select all the cash flows from the Net Cash Flow column including the initial expenditure (E7:E12). It is not essential to enter a guess but it may work more quickly if you do. The default is 10%. In this example the IRR should work out at 11.199%.

If the NPV is positive the IRR will be greater than the target rate and vice versa. If the IRR equals the target rate the NPV will be zero.

If the IRR function is entered in cell D25 and the NPV function in D24 the final result will appear as follows:

Excel for Surveyors

	A	B	C	D	E	F	G
1		Purchase price		1,250,000			
2		Rent		100,000			
3		Sale price		1,500,000			
4		NPV Rate		10.000%			
5							
6	Year	Purchase price	Rent	Sale price	Net Cash flow	PV factor	PV cash flow
7	0	-1,250,000			-1,250,000	1.0000	-1,250,000
8	1		100,000		100,000	0.9091	90,909
9	2		100,000		100,000	0.8264	82,645
10	3		100,000		100,000	0.7513	75,131
11	4		100,000		100,000	0.6830	68,301
12	5		100,000	1,500,000	1,600,000	0.6209	993,474
13							60,461
21							
22							
23							
24		**Net Present Value**		**60,461**			
25		**IRR**		**11.199%**			
26							

9.3 "What if" analysis

The beauty of Excel is that it allows you to change your mind about the various assumptions. You can change any of the initial data in D1:D5 at will, and you can see the effect on the NPV and the IRR instantly. Try changing some and see.

9.4 Rent quarterly in advance

Rent is frequently paid quarterly in advance. Because it is in advance you will receive your first payment of rent at the start of period 0. Annual inputs and outputs and rates of interest will need to be adjusted accordingly.

Example

Assume the facts of the previous example, except that rent is received quarterly in advance. We therefore divide the rent by 4. The target rate however is less than a quarter of the annual figure because of the effect of compounding. To calculate this we use the formula $(1+i)^{.25}-1$. In this case the quarterly effective rate corresponding to an annual effective rate of 10% is 2.4114%.

The corresponding Excel formula is:

=((1+i)^.25)–1

Our cash flow should look like this:

	A	B	C	D	E	F	G
1		Purchase price		1,250,000			
2		Rent per quarter		25,000			
3		Sale price		1,500,000			
4		NPV Rate		10.000%	per annum		
5							
6	Period	Purchase price	Rent	Sale price	Net Cash flow	PV factor	PV cash flow
7	0.00	-1,250,000	25,000		-1,225,000	1.0000	-1,225,000
8	0.25		25,000		25,000	0.9765	24,411
9	0.50		25,000		25,000	0.9535	23,837
10	0.75		25,000		25,000	0.9310	23,275
11	1.00		25,000		25,000	0.9091	22,727
12	1.25		25,000		25,000	0.8877	22,192
13	1.50		25,000		25,000	0.8668	21,670
21	3.50		25,000		25,000	0.7164	17,909
22	3.75		25,000		25,000	0.6995	17,487
23	4.00		25,000		25,000	0.6830	17,075
24	4.25		25,000		25,000	0.6669	16,673
25	4.50		25,000		25,000	0.6512	16,281
26	4.75		25,000		25,000	0.6359	15,897
27	5.00			1,500,000	1,500,000	0.6209	931,382
28							83,871
29							
30							
31		**Net Present Value**		**83,871**			
32		**IRR**		**11.743%**			
33							

Note that the IRR rate calculated is the quarterly rate received from the cash flow E7:E27and must therefore be adjusted to an annual basis. The formula in cell D32 is therefore:

=(1+IRR(E7:E27))^4–1

The NPV and IRR are greater because rent arrives earlier and more frequently. The sooner the rent arrives the less it is discounted and this has a positive effect on the results.

Cash flows running for several years on a monthly basis may become very long and can be difficult to manage. You can compress the cash flow, by hiding rows. When you have completed your cash flow select the rows you want to hide then select **Menu ⇒ Format ⇒ Row ⇒ Hide**. Calculations are unaffected. To restore rows select the rows either side of the hidden rows then select **Menu ⇒ Format ⇒ Row ⇒ Un-hide**.

9.5 Gearing

In our previous example it is possible that the investor borrowed money to finance the investment. This can incorporated into your cash flow.

Example

Again value the investor's interest in a freehold property. Rent is as before at £100,000 pa paid annually in arrears with five-year reviews. He intends to hold the property for five years. The target rate is 10% and growth is expected at 3%. However this time he decides to gear his purchase by 75%, on an interest-only loan.

There are now three more assumptions:

(a) the interest rate on the borrowings;
(b) the loan to value ratio (LTV);
(c) the tax rate.

The tax rate is assumed to be the corporate tax rate. Since it is possible to reclaim tax on interest payments we should include this benefit in our cash flow.

We set out the cash flow as in our first example, adding four new columns for Loan, Interest, Tax allowance, and cash flow after gearing.

Our cash flow should look like this:

Excel for Surveyors

	A	B	C	D	E	F	G	H	I	J	K
1		Purchase price		1,250,000		Loan		75%			
2		Rent p.a.		100,000		Interest		7.50%			
3		Sale price		1,500,000		Tax rate		30%			
4		NPV Rate		10.000%							
5											
6											
7	Year	Purchase price	Rent	Sale price	Net Cash flow	Loan, 75%	Interest @ 7.5%	Tax	Cash flow after gearing	PV factor	PV cash flow
8	0	-1,250,000	0		-1,250,000	937,500			-312,500	1.0000	-312,500
9	1		100,000		100,000		-70,313	21,094	50,781	0.9091	46,165
10	2		100,000		100,000		-70,313	21,094	50,781	0.8264	41,968
11	3		100,000		100,000		-70,313	21,094	50,781	0.7513	38,153
12	4		100,000		100,000		-70,313	21,094	50,781	0.6830	34,684
13	5		100,000	1,500,000	1,600,000	-937,500	-70,313	21,094	613,281	0.6209	380,799
14											229,269
15											
16		**Net Present Value**		**229,269**							
17		**IRR**		**25.840%**							

The new columns depend on the new information specified before our cash flow. The loan is quite simple. It is the purchase price multiplied by the LTV. It is a positive figure because it is money we receive at the beginning of the cash flow. On an interest only loan though we must remember to pay it back. In period 5 we need to subtract the loan.

Interest is assumed fixed at 7.50% of the loan payable in arrears. The annual interest in G9 is H2*F8 expressed as a negative flow, and is copied down for each year.

The tax allowance is an income. We include an extra column for this and multiply the interest by the corporate tax rate and express it as a positive cash flow.

Column I is the sum of the cash flows after allowing for the extra outflows and inflows as a result of the loan.

Finally, we determine the effect on worth of the loan.

Finding the NPV and IRR of the property on the 25% equity put into the property is exactly the same procedure as step 7 but using the "Cash flow after gearing" column instead of the Net Cash Flow column.

By gearing the investment with a 75% loan the NPV and IRR of the investment are increased but are now subject to greater risk because any failure of the investment to grow by 3% pa will result in a much greater proportionate decrease in the net return.

You should experiment with changes in the interest rates to see the effect it has on your results.

Other costs and receipts can be dealt with in the same way. The net cash flow rarely just consists of items we have considered. There will be a review and management fees, and maybe marketing costs. To include these, insert an extra column for each additional item.

9.6 Leaseholds

The cash flow approach is particularly useful when valuing leasehold interests sublet on a reviewable lease when the head rent is fixed.

Consider the following problem.

Example

The head rent payable for a lease with 10 years unexpired is £75,000 pa. It has just been sublet at the current rental value of £100,000 reviewable after five years. The initial profit rent is £25,000.

The normal practice of applying a multiplier adjusted for the risk and deteriorating nature of a leasehold does not accurately reflect growth in the rent receivable.

Before showing a solution to this problem we consider two identical profit rents:

	Rent received	Rent paid	Profit rent
Investment 1	£40,000	£15,000	£25,000
Investment 2	£100,000	£75,000	£25,000

Because of the gearing effect, in five years time these two profit rents will not have grown by the same amount:

	Rent in 5 years @ 3% growth	Rent paid	Profit rent
Investment 1	£46,370	£15,000	£31,370
Investment 2	£115,927	£75,000	£40,927

To incorporate this into the cash flow we add a row titled "Head Rent" and our cash flow will look like this:

Excel for Surveyors

	A	B	C	D	E	F	G
1			Investment 1			Investment 2	
2	Year	Income	Head rent	Profit rent	Income	Head rent	Profit rent
3	1	40,000	15,000	25,000	100,000	75,000	25,000
4	2	40,000	15,000	25,000	100,000	75,000	25,000
5	3	40,000	15,000	25,000	100,000	75,000	25,000
6	4	40,000	15,000	25,000	100,000	75,000	25,000
7	5	40,000	15,000	25,000	100,000	75,000	25,000
8	6	46,370	15,000	31,370	115,927	75,000	40,927
9	7	46,370	15,000	31,370	115,927	75,000	40,927
10	8	46,370	15,000	31,370	115,927	75,000	40,927
11	9	46,370	15,000	31,370	115,927	75,000	40,927
12	10	46,370	15,000	31,370	115,927	75,000	40,927
13							
14		**NPV**	**15.%**	**136,086**			**152,013**

Assuming a discount rate of 15% the NPV of Investment 1 is £136,086 and of Investment 2 is £152,013.

9.7 Valuations of over-rented property

In the late 1980s rental values of commercial properties fell below the rent actually being paid, leaving the freeholder with the prospect of no increase in rent at the following review, and in the case of a lease renewal the possibility of a reduction in rent. Conversely the lessee was faced with paying more for the property than competitors taking a new lease of similar property.

Several methods have been proposed[22] for dealing with this problem, and we consider one of these – the "Short-Cut DCF Method", broadly following Crosby's approach but with some modifications.

The principle is to assume that if the lease has an "upwards only" rent review clause, the freeholder will continue to receive the current income until there is a review when the rental value exceeds the current income. Like all situations regarding forecasting the future, certainty is impossible, and we can only base our forecast on an analysis of current data. We make the assumption that, even though values have sharply declined, the market is anticipating long-term future growth.

It may be helpful to consider this in graphical terms. In the chart below the rent for the next seven years is fixed at £25,000. However the rental value is only £15,000 and even allowing for growth over the next seven years at the projected rate, it will still be below the current rent. Therefore, assuming upwards-only terms, the rent will remain at £25,000 for the following seven years. At the following review the rental value is assumed to have risen above £25,000 and the next review/new lease will be at the full rental value at that time.

[22] See Crosby, F.N. [1992–2] *Journal of Property Investment and Valuation* 517–524.

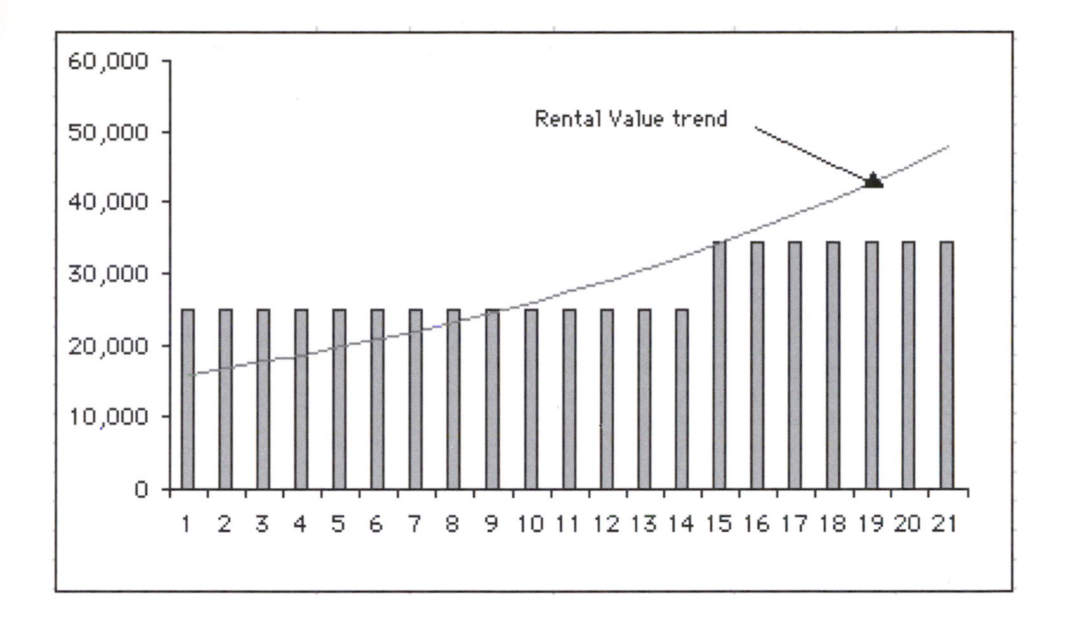

The DCF approach to the problem takes as its starting point an estimation of the implied growth rate based on the equated yield, the all-risks yield, and the rent review pattern (usually these days five-yearly). Crosby's approach is to use three factors obtained from standard valuation tables, but the same result can be achieved directly by the use of the formula:

$$g = \sqrt[n]{1 + (1 - \frac{k}{r})*((1+r)^n - 1)} - 1$$

where

g = the annual implied growth rate
k = the all-risks yield
r = the equated yield
n = the number of years in the standard rent review pattern for that class of property.

The worksheet equivalent of this formula is:

=(1+(1–B3/B2)*((1+B2)^B4–1))^(1/B4)–1

The assumptions on which this formula is based are as follows :

(a) The rents of other properties being analysed are at full rental value on a standard 5-year review pattern.
(b) The calculation is done at the start of a lease or rent review.
(c) The annual rate of increase in rental values is constant.
(d) Rental growth will continue in perpetuity.

In this case, taking the facts from the Crosby example, we have

$$g = \sqrt[5]{1 + (1 - \frac{.07}{.12})*(1.12^5 - 1)} - 1 = 5.67\%$$

The next consideration is whether the rental value will have risen above the current rent paid at the time of the next review. If it has, the freeholder is in a position to increase the rent to the full market rental value then; if not the rent will remain at its previous level. This can conveniently be established by creating a table of increases in rental values. Finally, we can set out the valuation assuming that the present rent remains unchanged until the rent has risen above it.

To set this up on a worksheet we proceed by entering the data. Below is a formula view of the worksheet.

	A	B	C	D	E	F
1			**Year**	**Increase**	**Rental value**	**Rent paid**
2	Equated yield	0.12	0	=(1+IG)^C2	15000	
3	All Risks yield	0.07	1	=(1+IG)^C3	=RV*D3	25000
4	Review Pattern	5	2	=(1+IG)^C4	=RV*D4	25000
5	Implied annual growth	=(1+(1-B	3	=(1+IG)^C5	=RV*D5	25000
6			4	=(1+IG)^C6	=RV*D6	25000
7	Rental value	15000	5	=(1+IG)^C7	=RV*D7	25000
8	Current rent	25000	6	=(1+IG)^C8	=RV*D8	25000
9			7	=(1+IG)^C9	=RV*D9	25000
10			8	=(1+IG)^C10	=RV*D1(	25000
11			9	=(1+IG)^C11	=RV*D1	25000
12			10	=(1+IG)^C12	=RV*D1:	25000
13			11	=(1+IG)^C13	=RV*D1:	25000
14			12	=(1+IG)^C14	=RV*D1<	25000
15			13	=(1+IG)^C15	=RV*D1!	25000
16			14	=(1+IG)^C16	=RV*D1(	25000
17			15	=(1+IG)^C17	=RV*D1	34315
18			16	=(1+IG)^C18	=RV*D1:	34315
19			17	=(1+IG)^C19	=RV*D1!	34315
20			18	=(1+IG)^C20	=RV*D2(	34315
21			19	=(1+IG)^C21	=RV*D2	34315
22			20	=(1+IG)^C22	=RV*D2:	34315
23			21	=(1+IG)^C23	=RV*D2:	34315
24						

The complete entry in cell B5 is =(1+(1–B3/B2)*((1+B2)^B4–1))^(1/B4)–1

Note that in constructing this it is only necessary to enter row 2 in columns D and E. Cells D3:E22 can be copied down from these as described earlier. In this case we have projected rental values for 21 years, but this need not be necessary in every case.

It is apparent that in year 7 the rental value is less than the rent actually received. (Note that in this case "Rental Value" refers to the current rental value on lease with five-yearly reviews.) Therefore the freeholder can expect the fixed rent for the following seven years.

We can now proceed with the valuation. This values the income for the first 14 years as a fixed income at the equated yield rate. On the reversion we make explicit allowance for growth up to that time, value the reversion at the all-risks yield of 7%, and then defer it to the present at the equated yield rate. Thus the first 14 years are a strict DCF valuation and the final reversion is of a rental value in perpetuity.

	A	B	C	D	E	F
1			Year	Increase	Rental value	Rent paid
2	Equated yield	12%	0	1.0000	15,000	
3	All Risks yield	7%	1	1.0567	15,851	25000
4	Review Pattern	5 years	2	1.1167	16,750	25000
5	Implied annual growth	5.67%	3	1.1800	17,700	25000
6			4	1.2469	18,704	25000
7	Rental value	15,000	5	1.3176	19,765	25000
8	Current rent	25,000	6	1.3924	20,886	25000
9			7	1.4714	22,070	25000
10			8	1.5548	23,322	25000
11			9	1.6430	24,645	25000
12			10	1.7362	26,043	25000
13			11	1.8347	27,520	25000
14			12	1.9387	29,081	25000
15			13	2.0487	30,730	25000
16			14	2.1649	32,473	25000
17			15	2.2877	34,315	34315
18			16	2.4174	36,261	34315
19			17	2.5545	38,318	34315
20			18	2.6994	40,491	34315
21			19	2.8525	42,788	34315
22			20	3.0143	45,215	34315
23			21	3.1853	47,779	34315
24						

The formula view of entries is as follows.

	H	I	J	K	L
1					
2					
3	Rent received			25000	
4	14	0.12		=(1-(1+I4)^-H4)/I4	=K3*K4
5					
6	Reversion to			15000	
7	Growth over 14 years		=D16		
8	YP perp	0.07	=1/I8		
9	14	0.12	=(1+I9)^-H9	=J7*J8*J9	=K6*K9
10					=SUM(L4:L9)
11					

The full entry in K4 is = 1–(1+I4)^–H4/I4

In normal view the final result is :

Excel to

	H	I	J	K	L
1					
2					
3	Rent received			25,000	
4	Years' Purchase 14 years	12%		6.6282	165,704
5					
6	Reversion to			15000	
7	Growth over 14 years		2.1649		
8	YP perp	7%	14.2857		
9	PV £1 14 years	12%	0.2046	6.3282	94,924
10					260,628

Chapter 10

Charts

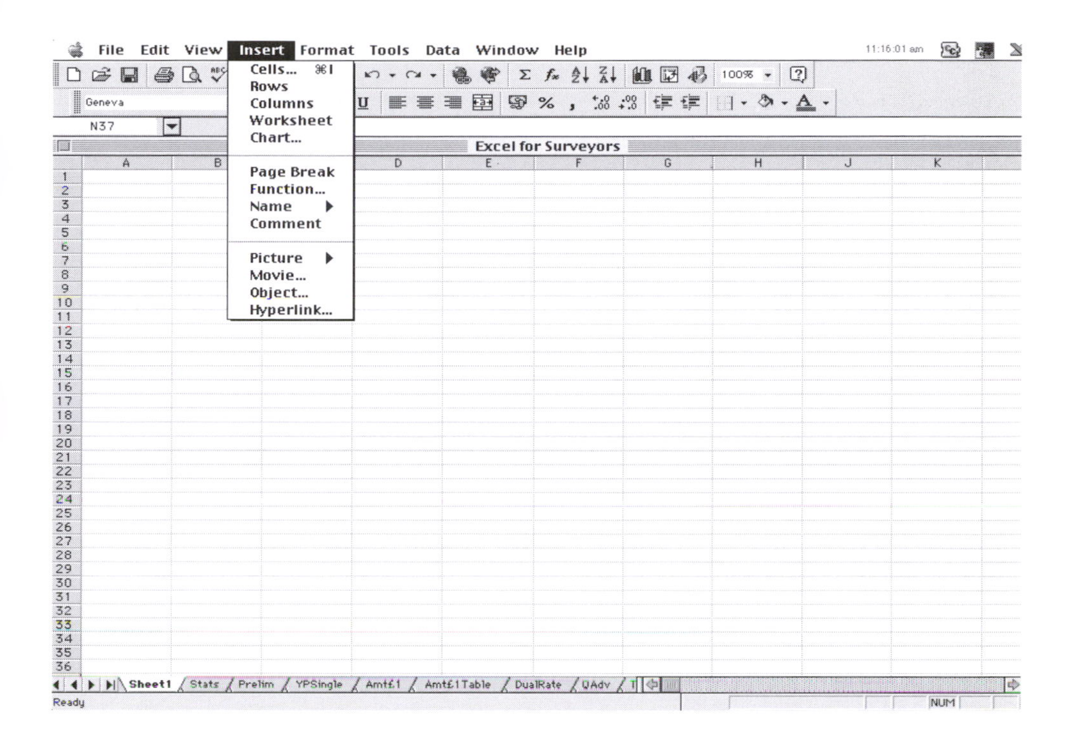

10.1 Types of chart in Excel

How many times have you heard the saying "a picture is worth a thousand words"? The graphical presentation of results is one of the most versatile and also one of the most informative means of conveying information to lay persons, and especially to clients.

An explanation of terminology is necessary here. "Graphics" in computer work normally refers to drawings, photographs and other similar objects which may be derived from scanned images of photographs "drawn" on screen with the drawing toolbar in Microsoft Office or other specialist applications. A "graph" on the other hand is a representation of a mathematical function such as $y=x^2$. Excel considers graphs under the general heading of "charts".

Charts may generally be divided into two groups – those which set out to make comparisons of sets of data, for example pie charts and bar charts, and those which describe visually a mathematical function, such as the graph of the Amount of £1.[22] Excel can produce both types but the procedures differ slightly. We shall

[22] A graph may be defined as a diagram (a series of one or more points, lines, line segments, curves, or areas) that represents the variation of a variable in comparison with that of one or more other variables.

refer to these as "charts" and "graphs" respectively for present purposes.

You will notice in the course of preparing charts and graphs that there are many forms available and you can experiment to find the one best suited to the purpose in hand. In Excel both are most easily produced with the Chart Wizard.

10.2 Setting up a chart

Example

You manage three blocks of flats on behalf of a tenants' association and have the data below. You wish to present this in graphical form to the annual meeting of the tenants.

We begin by entering headings into B1:D1 and A2:A6 and the data into B2:D6 as shown.

	A	B	C	D
1		Block A	Block B	Block C
2	Service charges	15,000	14,000	18,500
3	Repairs	7,500	8,500	5,000
4	Sinking fund	3,000	3,000	3,000
5	Management costs	1,500	1,200	1,850
6	Surplus	3,000	1,300	8,650
7				

(a) Select cells B2:D6 by dragging.

(b) Among the buttons on the toolbar at the top of the screen is the Chart Wizard. Click on this and the cursor will become a cross-hair.

(c) Drag to create a rectangle of convenient size and position to contain the chart. Step 1 of the Chart Wizard will appear as shown opposite.

(d) Step 1 asks for the chart type. Click on **Column** and **Next**.

(e) Step 2 You have already selected the range so click **Next**.

(f) Step 3 Click on **Menu** ⇒ **Format** ⇒ **Number** and **Next**.

(g) Step 4 shows a preview of the graph and suggests that the first column is the list of income and expenditure and the first row is the legend text (names of the blocks). Click Next.

(h) Step 5 asks for a Chart Title and titles for the two axes (x = horizontal and y = vertical). Complete these with whatever you wish to label the chart and you are done. The size of the chart can still be adjusted by clicking on it and dragging the handles, and the entire chart can be dragged to another position on the worksheet.

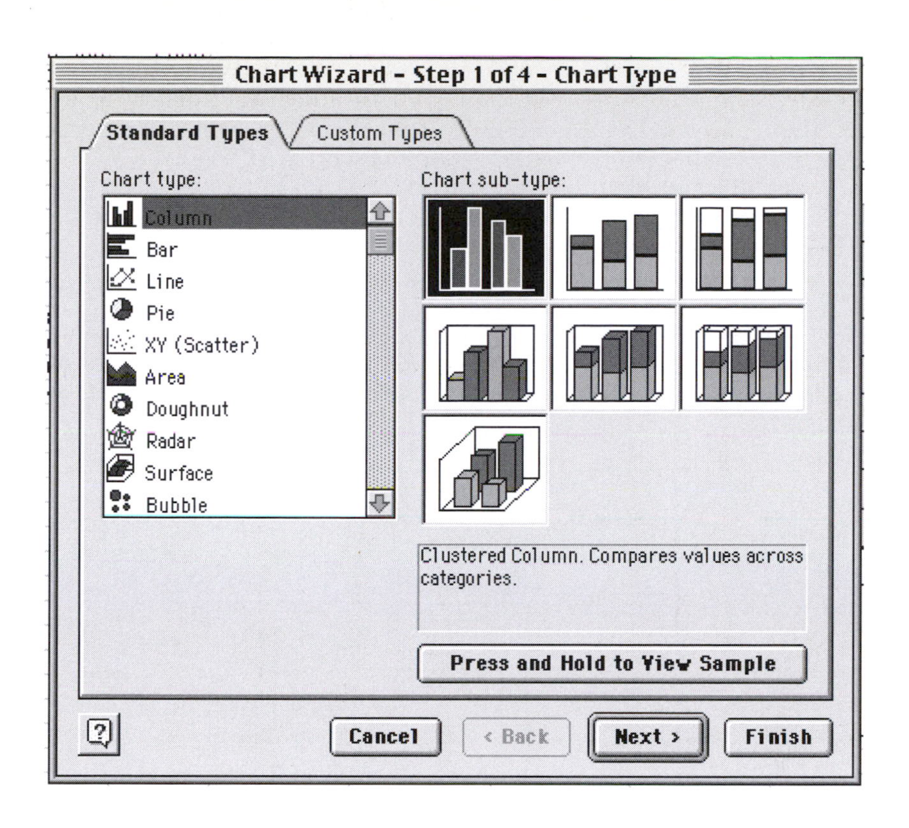
Chart Wizard - Step 1 of 4 - Chart Type
Standard Types
Custom Types
Chart type:
Column
Bar
Line
Pie
XY (Scatter)
Area
Doughnut
Radar
Surface
Bubble
Chart sub-type:
Clustered Column. Compares values across categories.
Press and Hold to View Sample
Cancel
< Back
Next >
Finish

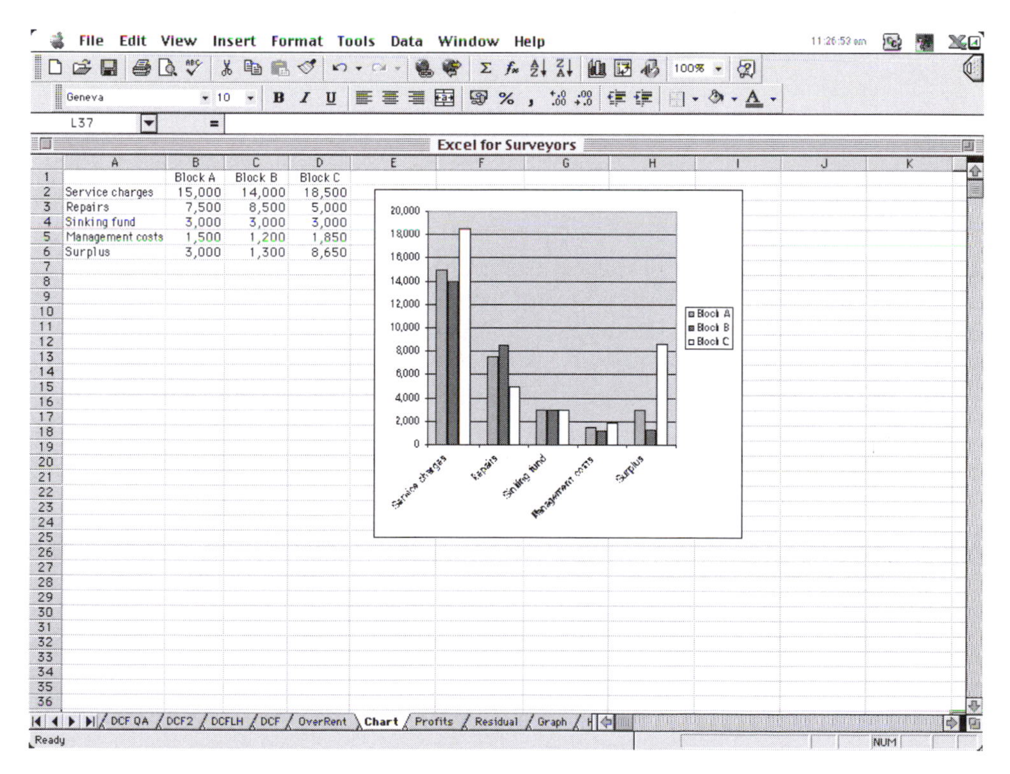
File Edit View Insert Format Tools Data Window Help
Excel for Surveyors
Block A
Block B
Block C
Service charges 15,000 14,000 18,500
Repairs 7,500 8,500 5,000
Sinking fund 3,000 3,000 3,000
Management costs 1,500 1,200 1,850
Surplus 3,000 1,300 8,650
DCF QA
DCF2
DCFLH
DCF
OverRent
Chart
Profits
Residual
Graph
Ready
NUM

10.3 Setting up a graph

We will set up the graph of the Amount of £1 and Present Value of £1 functions. The principle of the Amount of £1 is of course that £1 is invested at a given annual rate of interest and each year interest is added to the existing principal, so that the total increases proportionately more each year. The Present Value of £1 is the reciprocal of the Amount of £1 but is based on the same assumption.

The procedure is as follows.

(a) Enter the data in the worksheet in the usual way. In this case we enter years 0– 10 in column A, Amount of £1 in column B, and Present Value of £1 in column C, each at a rate of 5%, using formulae as previously discussed in chapter 6.

	A	B	C
1	Interest rate		5%
2	Years		
3	0	1.0000	1.0000
4	1	1.0500	0.9524
5	2	1.1025	0.9070
6	3	1.1576	0.8638
7	4	1.2155	0.8227
8	5	1.2763	0.7835
9	6	1.3401	0.7462
10	7	1.4071	0.7107
11	8	1.4775	0.6768
12	9	1.5513	0.6446
13	10	1.6289	0.6139

(b) Click on the Chart Wizard button and in Step 1 select Chart Type Line and the first subtype (displays trend over time or categories). Click **Next**.

(c) We will now see the draft graph. In Step 2 under the tab Data Range select the data, B3:C13. Do not include column A in this selection. Then select tab Series, click in Category (X) Axis labels and drag over A3:A13. Click **Next**.

(d) In Step 3 we have an opportunity to enter labels for the title and names for the X and Y axes. Enter "Amount and Present Value", "Years" and "Pounds" respectively, and click **Next**.

(e) Step 4 gives an option to place the graph in a new worksheet, but leave this and click **Finish**.

This is not quite what we need because Excel is still treating the year numbers as categories and the graph starts in the centre of the first category instead of on the Y axis.

(f) Finally click on the X-axis to select it and then **Menu** ⇒ **Format** ⇒ **Selected Axis**. Select the **Scale** tab and click on **Value (Y) axis crosses between categories** to remove the tick which is the default. The graph will now start at the Y-axis as in a standard graph.

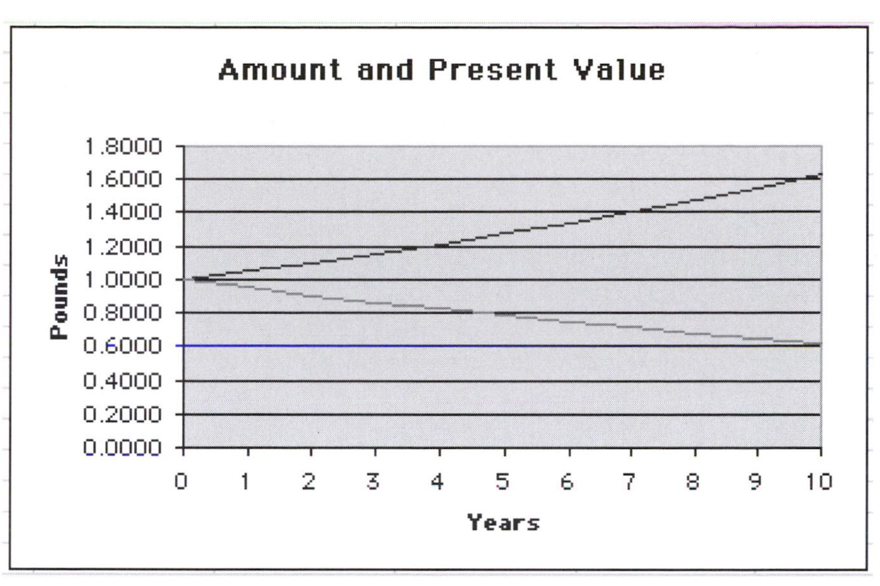

There are numerous other options for changing the appearance of your graph, and again we leave you to experiment with these.

10.4 Printing your chart

You can print the chart as a part of the worksheet by simply printing it as it stands, but before doing so ensure that the chart itself is not selected by clicking on another cell on the worksheet.

Alternatively, you may wish to print the chart as a separate sheet. In this latter case select the chart by clicking on it and then use the Print command (or the Print Preview command if you would like to see the result before printing).

10.5 Using charts for analysis

Example

You have an estate agency practice and have recorded information about the average sale price of similar semi-detached houses on a particular estate during past years. You wish to discover whether in fact there has been an underlying upwards trend in the values, bearing in mind the collapse of the market at the end of the 1980s, and to illustrate this on a graph.

This requires some explanation of an Array and the use of the GROWTH function from the Insert menu.

An Array permits us to make multiple calculations by entering one formula only, and can be used for many purposes. You can find more information about arrays in the Help Menu. We will confine ourselves to this one particular example.

The steps are:

(a) Head columns A1 B1 and C1 Year, Price and Trend respectively.
(b) Enter the data into consecutive columns of the worksheet in columns 1 and 2 (cells A2 to B16).

	A	B	C
1	Year	Price	Trend
2	1985	80,000	
3	1986	82,500	
4	1987	85,000	
5	1988	80,000	
6	1989	75,000	
7	1990	71,000	
8	1991	74,000	
9	1992	79,000	
10	1993	83,000	
11	1994	83,000	
12	1995	84,000	
13	1996	85,000	
14	1997	87,000	
15	1998	89,000	
16	1999	91,000	

(c) Select cell C2.

(d) Click on **Menu** ⇒ **Insert** ⇒ **Function** and Step 1 of the Function Wizard will appear. From this select **Function Category Statistical** and **Function Name Growth**. Click **OK**.

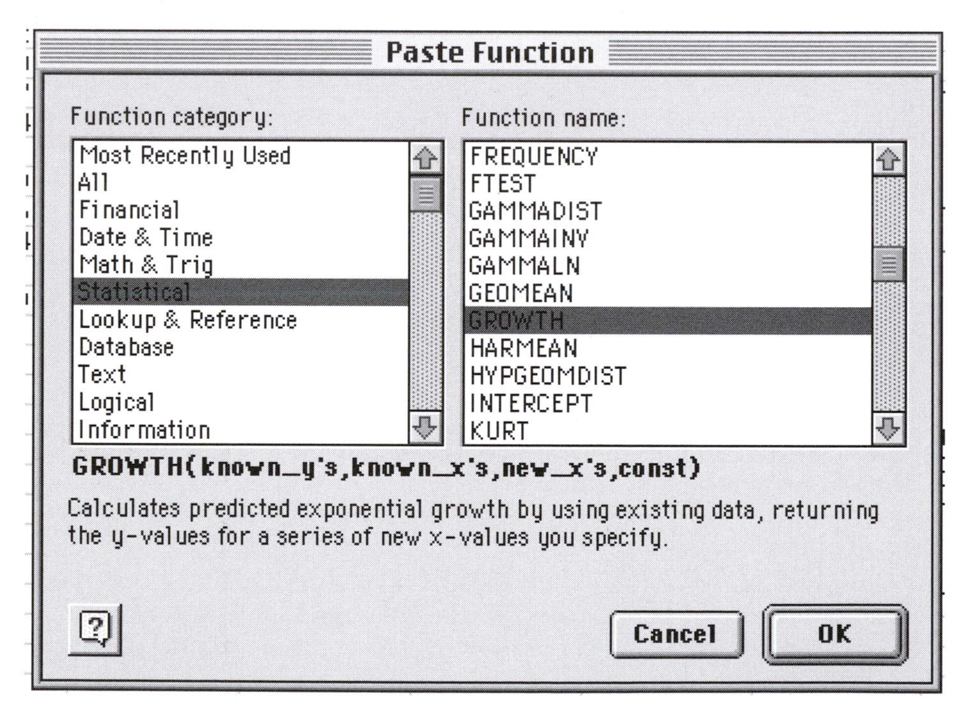

(e) In Step 2 of the Function Wizard select cells B2:B16. Disregard the other entries at this step in this case and click **Finish**. The first projected value of the trend will appear.

(f) Select cell C2: and drag down to cell C16, but do not hit the **Return** key.

(g) Click in the formula bar as if you were going to edit the formula, but then press **Control–Shift–Return**.[24] This will turn the column into an array and fill it with the remaining trend values.

	A	B	C
1	Year	Price	Trend
2	1985	80,000	76,803
3	1986	82,500	77,487
4	1987	85,000	78,177
5	1988	80,000	78,874
6	1989	75,000	79,576
7	1990	71,000	80,285
8	1991	74,000	81,000
9	1992	79,000	81,721
10	1993	83,000	82,449
11	1994	83,000	83,184
12	1995	84,000	83,924
13	1996	85,000	84,672
14	1997	87,000	85,426
15	1998	89,000	86,187
16	1999	91,000	86,954

We can now proceed to plot the graph of the trend with the Chart Wizard as before. Select A2:C16 so as to include the column titles, and drag across blank cells to create the chart. At steps 2 and 3 it is probably most appropriate to select a line chart, but you can try others as you wish. At Step 4 ensure that column 1 is selected for category X labels and row 1 for the Y values. Step 5 asks for the Chart Title, axes titles and whether you want the legend.

The Y-axis will default to a base of zero, but this can be changed to emphasise the trend and then **Menu ⇒ Format ⇒ Selected Axis**. Click on the Scale tab and in the "Category (X) Axis crosses at" box enter 70000.

The final result appears as:

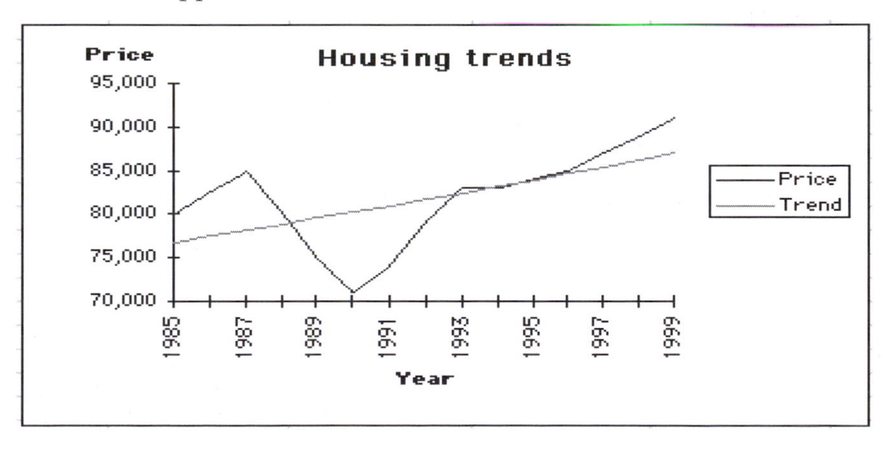

[24] Command–Return on the Macintosh.

You can modify the appearance of your chart by double-clicking on the chart itself and then on any part of it, for example on an axis if you want to change the formatting of the numbers or on a line to change the colour. We leave you to experiment with this.

Chapter 11

More valuations

In this chapter we look at some further examples of the way in which Excel can be set up to do valuations. You are recommended to set these out on your own computer for practice, and because a library of standard valuations can be very useful. It is frequently easier to take an existing example when you have a similar problem, and modify it instead of setting one up from scratch.

The term "Residual Valuation" is usually applied to problems involving development, but in its broadest sense relates to any valuation which consists of an estimate of the value of a property after allowing for various items of expenditure. Thus a Profits Valuation where the object is to find the rental bid after allowing for tenants' outgoings may also be regarded as a residual.

Setting up a residual valuation in Excel will follow the same general layout as a conventional calculation set out on paper. However the many format options allow for it to be tailored to the particular valuation under consideration.

11.1 Residual valuations

The general basis of a residual valuation is:

Site value = Value on completion – Cost of work – Profit

Alternatively an appraisal study may be expressed as:

Profit = Value on completion – Site Cost – Cost of work

Either way, the calculation may be set up on a worksheet to enable "What if" experiments to be done.

Example

For this example we take the valuation set out in *Valuation and Development Appraisal*.[25] Briefly, this assumes a site with outline planning permission for 500 m^2 of standard unit shops with 5,000 m^2 of offices above. The valuation, (slightly rearranged and abbreviated) is set out over.

Frequently some of the inputs, in this case the finance rate, will be the same in different parts of the calculation, and it is therefore convenient to use a single cell for this data and to copy it as required to other cells. Then, for example, to change the finance rate from 14% to 15%, the result can be recalculated in one operation.

---

[25] Clive Darlow *Valuation and Development Appraisal* 2nd ed 1988, *Estates Gazette* pp13–14.

VALUE OF SCHEME		
Income		
Shops – 500 m² less 10% @ £130/m²	58,500	
Offices – 5,000 m² @ £150/m²	<u>600,000</u>	
Total income	658,500	
Yield @ say 7%	<u>14.2875</u>	
Capital value		9,407,000
COST OF SCHEME		
Building costs		
Shops – 500 m² gross @ £350/m²	175,000	
Offices – 5,000 m² gross @ £750/m²	<u>3,750,000</u>	
	3,925,000	
Ancillary costs		
Access roads etc	<u>200,000</u>	
	4,125,000	
Professional fees, 12.5%	<u>515.500</u>	
	4,640,500	
Contingencies @ 3% of total costs incurred	<u>139,000</u>	
	4,779,500	
Short term finance @ 14%		
On building etc costs for 1/2 period	493,500	
	<u>5,273,000</u>	
on total costs incurred on completion of building work for full length of letting delay (ie 6 months)	<u>357,000</u>	
	5,630,000	
Letting and Sale Fees, 15% of income	99,000	
Advertising and marketing	25,000	
Fees for selling commercial investment @ 2%	188,000	
	<u>5,942,000</u>	
Return for risk and profit 17% of capital value	1,599,000	
TOTAL EXPECTED COSTS		<u>7,541,000</u>
Site value in 2.5 years		1,866,000
PV £1, 2.5 years, 14%		<u>0.72067</u>
		1,345,000
less acquisition costs, 2.5%		<u>33,500</u>
SITE VALUE TODAY		<u>1,311,500</u>

Entered on the worksheet, the formula view is as follows.

Note that the figures differ slightly from those in *Valuation and Development Appraisal* because no rounding is done.

Excel for Surveyors

	A	B	C	D	E
1	Discount rate	**0.14**			
2					
3	**VALUE OF SCHEME**				
4	*Income*				
5	-shops – 450m2	**450**	**130**	=B5*C5	
6	-offices 4,000m2 net	**4000**	**150**	=B6*C6	
7	Total income			=SUM(D5:D6)	
8	Yield @ say	**0.07**		=1/B8	
9	Capital value				=D7*D8
10	**COST OF SCHEME**				
11	*Building costs*				
12	Shops – 500m2 gross @ £350m2	**500**	**350**	=B12*C12	
13	Offices – 5,000m2 gross @ £750m2	**5000**	**750**	=B13*C13	
14				=SUM(D12:D13)	
15	*Ancillary Costs*				
16	Access road, services, etc, say			**200000**	
17				=D14+D16	
18	Professional fees	**0.125**		=(D14+D16)*B18	
19	*Short-term finance*			=SUM(D17:D18)	
20	Contingencies @ 3% total costs	**0.03**		=D19*B20	
21				=D19+D20	
22	building costs etc for 1/2 period	=B1		=D21*((1.14)^0.75-1)	
23				=D21+D22	
24	for 3 months' letting delay	=B1		=D23*((1+B24)^0.5-1)	
25				=D23+D24	
26	Letting and Sale Fees % of income	**0.15**	=B26*D7		
27	Advertising and marketing		**25000**		
28	Fees for selling comercial investment @	**0.02**	=E9*B28	=SUM(C26:C28)	
29				=D25+D28	
30	Return for risk and profit % of capital value	**0.17**		=E9*B30	
31	TOTAL EXPECTED COSTS				=D29+D30
32	Site value on completion				=E9-E31
33	PV £1	=B1	**2.5**		=(1+B33)^-C33
34					=E32*E33
35	Less acquisition costs	**0.025**			=E34-E36
36	**SITE VALUE TODAY**				=E34*0.975
37					

The completed worksheet is shown above.

Note that those cells that are shown in Bold are for data entry and the remainder are all based on formulae derived from the data. It is now possible to do "what if" and scenarios as discussed previously.

If you are not sure where some of the calculated values are derived from, set up this on your worksheet and use **Menu** ⇒ **Tools** ⇒ **Audit** which we discussed in chapter 8.2.

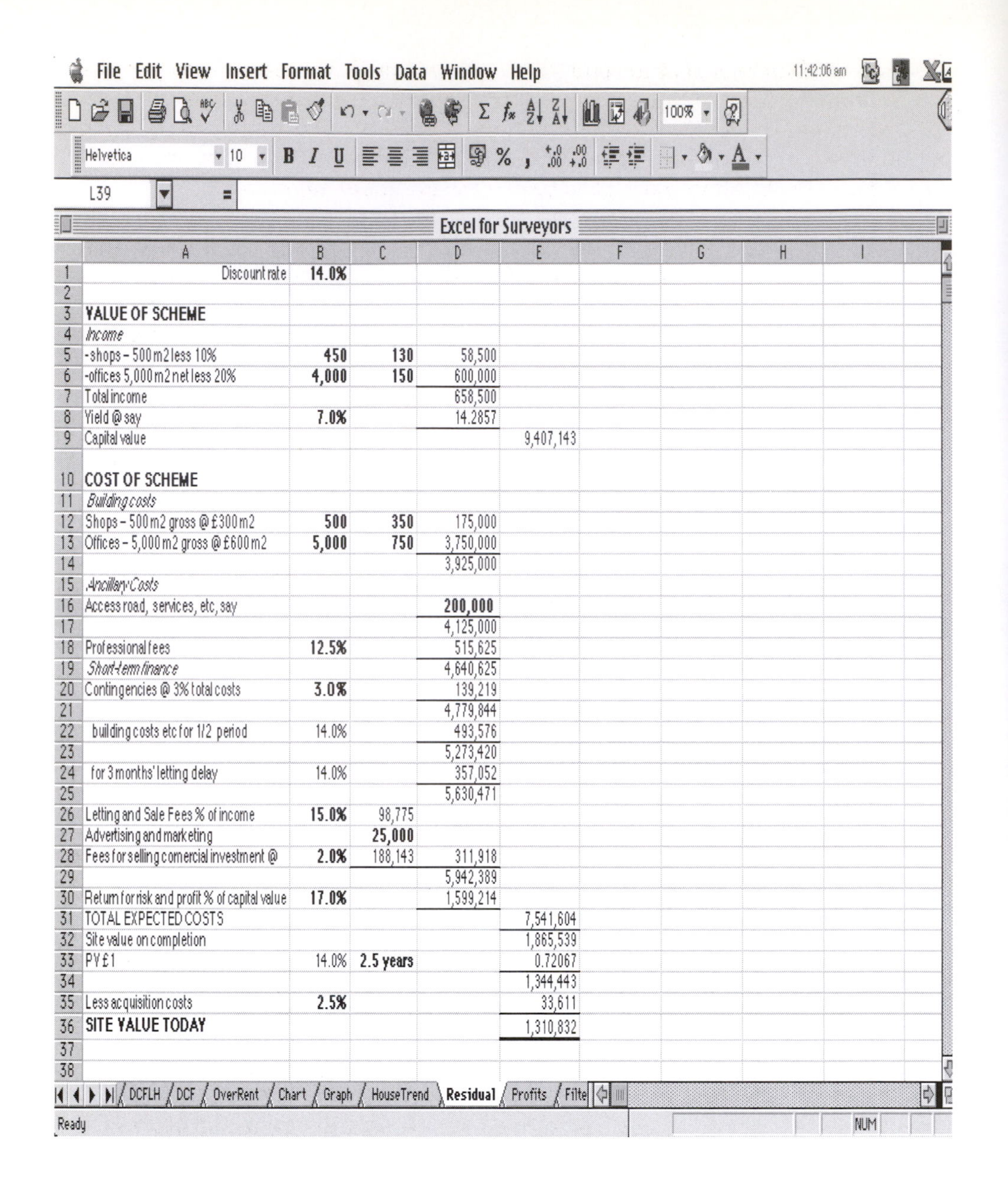

Excel for Surveyors

	A	B	C	D	E
1	Discount rate	**14.0%**			
2					
3	**VALUE OF SCHEME**				
4	*Income*				
5	-shops - 500m2 less 10%	**450**	**130**	58,500	
6	-offices 5,000m2 net less 20%	**4,000**	**150**	600,000	
7	Total income			658,500	
8	Yield @ say	**7.0%**		14.2857	
9	Capital value				9,407,143
10	**COST OF SCHEME**				
11	*Building costs*				
12	Shops - 500m2 gross @ £300m2	**500**	**350**	175,000	
13	Offices - 5,000m2 gross @ £600m2	**5,000**	**750**	3,750,000	
14				3,925,000	
15	*Ancillary Costs*				
16	Access road, services, etc, say			**200,000**	
17				4,125,000	
18	Professional fees	**12.5%**		515,625	
19	*Short-term finance*			4,640,625	
20	Contingencies @ 3% total costs	**3.0%**		139,219	
21				4,779,844	
22	building costs etc for 1/2 period	14.0%		493,576	
23				5,273,420	
24	for 3 months' letting delay	14.0%		357,052	
25				5,630,471	
26	Letting and Sale Fees % of income	**15.0%**	98,775		
27	Advertising and marketing		**25,000**		
28	Fees for selling comercial investment @	**2.0%**	188,143	311,918	
29				5,942,389	
30	Return for risk and profit % of capital value	**17.0%**		1,599,214	
31	TOTAL EXPECTED COSTS				7,541,604
32	Site value on completion				1,865,539
33	PV £1	14.0%	**2.5 years**		0.72067
34					1,344,443
35	Less acquisition costs	**2.5%**			33,611
36	**SITE VALUE TODAY**				1,310,832
37					
38					

11.2 Construction costs

Excel is perfect for mimicking the S-curve pattern that is typical of construction expenditure during a development. Construction costs throughout the development period are rarely uniform as they may be depicted in some cash flows (see section 11.3). The typical pattern of construction expenditure consists of smaller amounts spent in the first few months while site clearance and footings are put in place. A large proportion of the total cost is spent in the middle of the construction period while material and labour are used up on the building.

Expenditure then tails of at the end when the finishing touches are being applied. If the cumulative expenditure is graphically represented an S-curve results.

To reproduce this in Excel a series of numbers are entered against each specific period to represent an apportionment of construction cost. For a 12-month period with a total construction cost of £500,000 these numbers might be:

Month	Apportionment	Weight	Cost in Month	Cumulative
0	1	0.04	20,833	20,833
1	1	0.04	20,833	41,666
2	2	0.08	41,667	83,333
3	2	0.08	41,667	125,000
4	3	0.13	62,500	187,500
5	3	0.13	62,500	250,000
6	3	0.13	62,500	312,500
7	3	0.13	62,500	375,000
8	2	0.08	41,667	416,667
9	2	0.08	41,667	458,333
10	1	0.04	20,833	479,167
11	1	0.04	20,833	500,000
12				
Total	24			

In formulas view this is:

Excel for Su

	A	B	C	D	E
1	Month	Apportionment	Weight	Cost in month	Cumulative
2	0	1	0.04	=500000*C2	=D2
3	1	1	0.04	=500000*C3	=D3+E2
4	2	2	0.08	=500000*C4	=D4+E3
5	3	2	0.08	=500000*C5	=D5+E4
6	4	3	0.13	=500000*C6	=D6+E5
7	5	3	0.13	=500000*C7	=D7+E6
8	6	3	0.13	=500000*C8	=D8+E7
9	7	3	0.13	=500000*C9	=D9+E8
10	8	2	0.08	=500000*C10	=D10+E9
11	9	2	0.08	=500000*C11	=D11+E10
12	10	1	0.04	=500000*C12	=D12+E11
13	11	1	0.04	=500000*C13	=D13+E12
14	12				
15	Total				

The apportionment figures are put in simply to allow the replication of the S-curve and the numbers in this example represent a simple pattern. These figures would be provided by the quantity surveyor at the more advanced stages of appraisal. The weight is then found by dividing each monthly figure by the total. The construction cost can then be found by multiplying this weight by the total anticipated construction cost. You can see in this column that expenditure is heaviest in the middle of the building period. To show the S-curve the cumulative cash flow must be found, as in the final column.

Excel for Su

	A	B	C	D	E
1	Month	Apportionment	Weight	Cost in month	Cumulative
2	0	1	0.04	20,000	20,000
3	1	1	0.04	20,000	40,000
4	2	2	0.08	40,000	80,000
5	3	2	0.08	40,000	120,000
6	4	3	0.13	65,000	185,000
7	5	3	0.13	65,000	250,000
8	6	3	0.13	65,000	315,000
9	7	3	0.13	65,000	380,000
10	8	2	0.08	40,000	420,000
11	9	2	0.08	40,000	460,000
12	10	1	0.04	20,000	480,000
13	11	1	0.04	20,000	500,000
14	12				
15	Total				

Highlight this column, click on the chart wizard, and select the first line graph option. By holding down the view button you will be able to see instantly the S-shape. You may then proceed to add in titles and labels as in the previous examples.

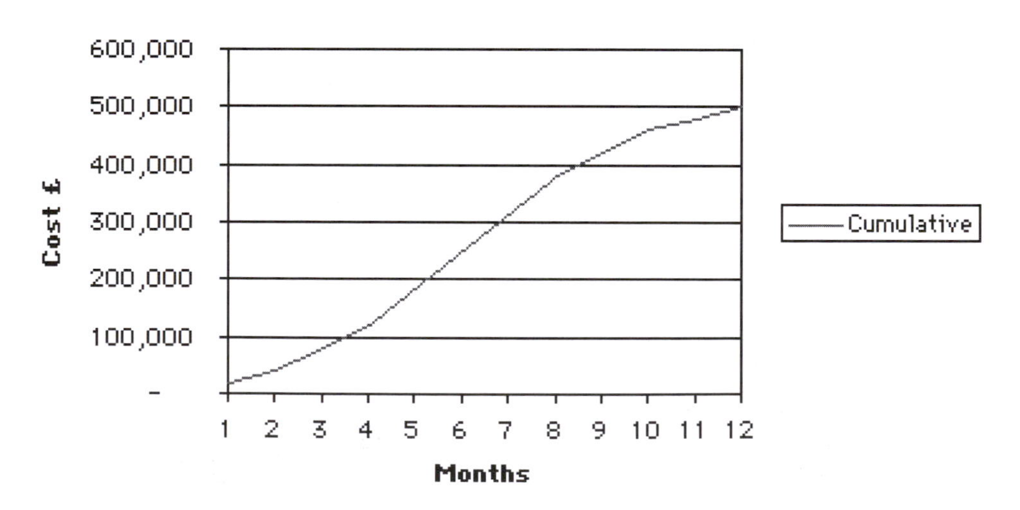

11.3 Profits valuations

A further example of setting up a worksheet for a valuation is given below. You may like to try setting this up yourself.

We follow previous examples here and show first the formula view of the cell entries, and then the computed results.

	A	B	C
1	**COLD COM**		
2			
3	Sales	**189000**	
4	Purchases	**86325**	=B3-B4
5			
6	*Less* Overheads		**61593**
7			=C4-C6
8	Adj for private expenses		**3300**
9	Net profit		=C7+C8
10			
11	Add :-		
12	Loan interest		**19000**
13			=C9+C12
14	*Less* :-		
15	Tenant's capital	**30000**	
16	**0.099**	=A16	=B15*B16
17	Adjusted Net Profit		=C13-C16
18	YP perp	**0.0675**	=1/B18
19	Capital Value		=C17*C18
20			

As before, cells shown in Bold are the original data and the remainder are calculated by the formulae.

	A	B	C
1	**COLD COMFORT HOTEL**		
2			
3	Sales	**189,000**	
4	Purchases	**86,325**	102,675
5			
6	*Less* Overheads		**61,593**
7			41,082
8	Adj for private expenses		**3,300**
9	Net profit		44,382
10			
11	Add :-		
12	Loan interest		**19,000**
13			63,382
14	*Less* :-		
15	Tenant's capital	**30,000**	
16	**@ 9.9%**	9.9%	2,970
17	Adjusted Net Profit		60,412
18	YP perp	**6.75%**	14.8148
19	Capital Value		894,993

Chapter 12

Using Excel for managing data

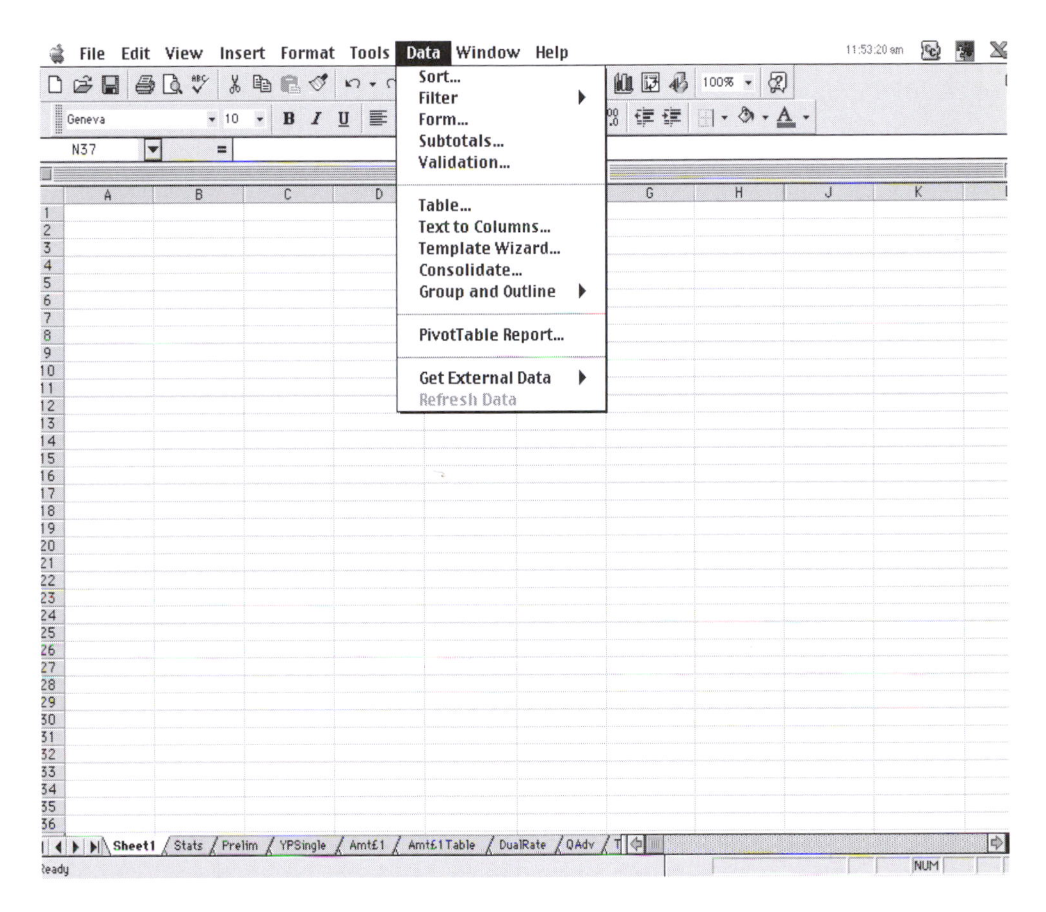

12.1 Databases

It will be apparent by now that Excel can store both data and instructions about what the computer is to do with that data (the formulae and the functions). So far we have concentrated on the use of the latter, but another use of Excel is to store data, sort it, classify and summarise it.

To be of practical use, the data must be stored in a systematic fashion. A standard telephone directory has a column of names and a column of numbers associated with those names, and would be of little use if the names and numbers were arranged in a random fashion.

Suppose we wish to store data about houses for sale in our area, and to be able to retrieve data about any particular house, houses in a given price range, or houses with more than three bedrooms. We start by listing the information we wish to record about each property, and using one column of our worksheet for each item of information. Each row can then record the details of one property. In database parlance the column is referred to as a "field" and the row as a "record". The first row of the sheet should be used to designate the name of the field, as this will be important later.

For our example we will designate the columns as follows :

Excel for Surveyors

	A	B	C	D	E	F	G	H	I	J	K
1											
2											
3	Ref	Num	Street	Price	Type	Age	Floors	Area	Recep	Bed	Garage

Basic data entry from sales etc is initially a matter of considerable spadework, but once done we can rapidly access any information with the **Sort** facility.

The worksheet below shows the data from recent sales of 73 houses.

Excel for Surveyors

	A	B	C	D	E	F	G	H	I	J	K
5	Ref	Num	Street	Price	Type	Age	Floors	Area	Recep	Bed	Garage
6	1	70	Badger Road	35,000	2	16	2	566	2	3	0
7	2	48	MouseWalk	35,000	4	60	2	545	1	1	1
8	3	88	The Leavens	36,000	3	25	2	610	2	3	1
9	4	26	Badger Road	37,500	4	25	2	645	2	3	1
10	5	3	Falcon Road	38,000	1	5	2	594	1	3	0
11	6	32	MouseWalk	39,000	4	80	2	451	1	2	1
12	7	88	The Leavens	39,500	4	30	2	395	1	2	1
13	8	18	Water Road	39,950	3	59	2	299	1	1	1
14	9	63	The Leavens	39,950	4	50	3	607	1	2	1
15	10	76	Silver Way	39,995	3	5	2	432	1	1	1
16	11	41	Water Road	40,000	4	40	2	753	2	2	1
17	12	84	Falcon Road	43,950	4	90	2	488	1	2	1
18	13	21	Falcon Road	43,950	2	20	2	365	1	1	0
19	14	85	The Leavens	44,950	2	3	2	690	2	3	0
20	15	25	The Leavens	45,000	4	80	2	746	2	3	1
21	16	48	Water Road	45,500	3	55	2	508	2	2	1
22	17	32	MouseWalk	47,950	4	85	2	551	2	2	2
23	18	40	Badger Road	48,950	1	20	2	660	1	2	0
24	19	95	Starling Road	51,950	3	90	2	440	1	2	1
25	20	54	Falcon Road	53,000	3	80	2	652	2	3	1
26	21	1	Badger Road	53,995	2	2	2	337	1	1	0
27	22	44	The Leavens	54,000	3	2	2	293	1	1	1
28	23	32	Falcon Road	54,495	3	60	2	568	1	2	1
29	24	22	Starling Road	54,500	3	50	2	761	2	3	1
30	25	81	Falcon Road	54,500	4	8	2	267	1	1	2
31	26	45	MouseWalk	55,000	4	80	2	520	1	2	2
32	27	76	Starling Road	56,950	1	0	2	320	1	2	0
33	28	82	The Leavens	57,000	4	80	2	624	2	2	2
34	29	58	The Leavens	57,995	2	55	2	650	2	2	0
35	30	48	Falcon Road	58,450	2	20	2	390	1	1	0
36	31	45	MouseWalk	58,950	2	60	2	640	1	3	0
37	32	22	Badger Road	59,950	1	0	2	462	2	2	0
38	33	57	MouseWalk	59,950	1	0	2	521	2	2	0
39	34	13	Water Road	59,950	4	80	2	455	2	2	2
40	35	95	Silver Way	59,950	2	20	2	700	2	3	1
41	36	99	Badger Road	61,950	3	5	2	289	1	1	1

Excel for Surveyor

	A	B	C	D	E	F	G	H	I	J	K
42	37	58	Mole Way	62,000	4	10	2	310	1	1	2
43	38	46	The Leavens	62,500	4	50	2	520	2	2	2
44	39	33	Silver Way	62,500	4	50	2	543	2	3	2
45	40	74	Mole Way	62,500	1	0	2	533	2	2	0
46	41	18	MouseWalk	64,950	3	0	2	451	1	1	1
47	42	54	Starling Road	64,950	2	40	2	588	2	2	1
48	43	31	Falcon Road	64,995	4	5	2	452	1	2	2
49	44	68	The Leavens	65,000	3	90	2	554	2	2	1
50	45	59	Mole Way	65,000	3	60	2	540	1	1	1
51	46	9	Falcon Road	65,000	4	9	2	478	1	2	2
52	47	74	Mole Way	65,000	2	10	2	730	2	3	1
53	48	85	Falcon Road	65,500	3	5	2	316	1	1	1
54	49	13	Starling Road	67,500	4	20	2	688	1	3	2
55	50	46	MouseWalk	67,500	4	30	2	646	1	2	2
56	51	22	Water Road	67,500	4	25	2	685	1	3	2
57	52	31	Starling Road	67,500	2	20	2	695	2	3	1
58	53	44	The Leavens	68,450	4	60	2	515	1	2	2
59	54	2	Starling Road	68,950	1	0	2	406	1	1	0
60	55	27	Badger Road	68,950	4	15	2	504	1	3	2
61	56	48	Mole Way	69,950	2	10	2	435	1	1	1
62	57	19	Water Road	71,000	4	5	2	310	1	1	2
63	58	97	Badger Road	71,500	4	6	2	368	1	1	2
64	59	44	Mole Way	73,500	2	5	2	379	1	1	1
65	60	29	Water Road	73,950	3	10	2	404	1	1	1
66	61	43	Mole Way	74,000	4	90	2	570	2	2	2
67	62	6	MouseWalk	75,000	2	55	2	659	2	3	1
68	63	5	Starling Road	75,950	2	60	2	604	1	2	1
69	64	71	Falcon Road	79,950	3	20	2	730	2	3	1
70	65	85	Falcon Road	80,950	4	90	2	591	1	3	2
71	66	60	MouseWalk	83,950	3	65	2	621	1	2	1
72	67	49	Mole Way	84,950	4	60	2	750	2	3	2
73	68	95	Badger Road	84,950	4	5	2	416	1	2	2
74	69	45	The Leavens	85,950	2	25	2	522	2	3	1
75	70	39	Silver Way	89,950	2	35	2	687	2	3	1
76	71	38	Falcon Road	93,000	3	15	2	800	2	3	1
77	72	61	Falcon Road	96,950	1	0	2	399	2	2	0
78	73	89	Water Road	109,950	2	55	2	660	1	3	1

Finding a particular property knowing its reference number is of course a simple matter if we use the **Menu** ⇒ **Edit** ⇒ **Find** command and enter the number. However **Find** will find all occurrences of the number, for example finding "49" will also find the price "64,950". To prevent this select the Num column by clicking on the column heading – this will restrict the search to the reference numbers.

We use the **Sort** and **Filter** commands to perform more complex searches.

12.2 Sorting

The data are already arranged in ascending order of price, so that it is easy to scan down the list for properties in a particular price range. However we may wish to arrange the properties in street order. We proceed as follows :

First, note that the street number is in column B and the street name in column C. This will enable us to sort by street name – otherwise properties would be sorted by street number and street names would be randomly arranged.

It will be much easier to refer to all the properties if we first give a name to the entire data set, including the column headings. Select cells A1:K78 and give these a name. We will refer to this as "houses".

To name the selected area of the sheet, go to **Menu** ⇒ **Insert** ⇒ **Name** ⇒ **Define**. The screen will show you any existing names in that workbook and ask you for a name for your selection. Type "houses" and **Return**.

Now use the **Menu** ⇒ **Edit** ⇒ **Goto** command and select "houses" for sorting and then **Menu** ⇒ **Data** ⇒ **Sort.** Another dialogue box will appear with three sets of criteria.

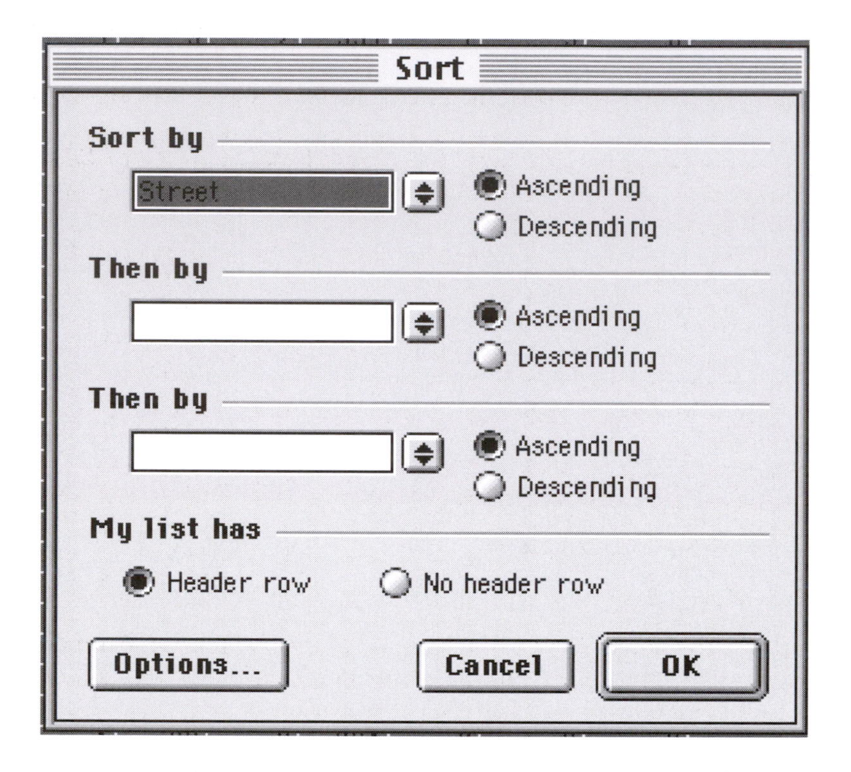

Click on the down arrow and a list of the column headings will appear. Choose "Street" and **Return**. Data will be sorted according to street name, but the street numbers will still be in random order.

To sort by street numbers as well, select **Menu** ⇒ **Data** ⇒ **Sort** again, but this time click on the second arrow and choose "Num" and **Return**. This time you will find that streets are arranged in alphabetical order, and within each street properties are numbered consecutively.

You will no doubt have noticed the "ascending" and "descending" buttons. If we had wished to sort streets in reverse alphabetical order or street numbers in descending order we would simply have clicked on the appropriate button.

In a large database one might of course use all three criteria.

The reason why we gave each property a reference number should by now be obvious. Apart from being able to find the property by its reference, we can easily restore the original order of data by sorting on column A.

12.3 Filtering

This is an alternative to simply sorting the data. By filtering we can effectively hide all records except those which we are interested in, and for this purpose we have several alternatives.

Suppose that we wish to see only details of properties in MouseWalk, we proceed as follows

(a) Select row 1.
(b) **Menu** ⇒ **Sort** ⇒ **Filter** ⇒ **Autofilter**. An arrow will appear at each column heading.
(c) Click and hold on the arrow on the Street heading. A list of streets will appear.

Excel for Surveyors

	A	B	C	D	E	F	G	H	I	J	K
1			(All)								
2			(Top 10...)								
3	**Ref**	**Num**	(Custom...)	**Price**	**Type**	**Age**	**Floors**	**Area**	**Recep**	**Bed**	**Garage**
4	1	70	Badger Road	35,000	2	16	2	566	2	3	0
5	2	48	Falcon Road	35,000	4	60	2	545	1	1	1
6	3	88	Mole Way	36,000	3	25	2	610	2	3	1
7	4	26	MouseWalk	37,500	4	25	2	645	2	3	1
8	5	3	Silver Way	38,000	1	5	2	594	1	3	0
9	6	32	Starling Road	39,000	4	80	2	451	1	2	1
10	7	88	Street	39,500	4	30	2	395	1	2	1
11	8	18	The Leavens	39,950	3	59	2	299	1	1	1
12	9	63	Water Road	39,950	4	50	3	607	1	2	1
13	10	76	(Blanks)	39,995	3	5	2	432	1	1	1
14	11	41	(NonBlanks)	40,000	4	40	2	753	2	2	1
15	12	84	Falcon Road	43,950	4	90	2	488	1	2	1
16	13	21	Falcon Road	43,950	2	20	2	365	1	1	0
17	14	85	The Leavens	44,950	2	3	2	690	2	3	0

(d) Move down the list to "MouseWalk" and release the mouse. All records will disappear except those relating to properties in MouseWalk. These can be sorted in the manner we have already discussed.

Excel for Surveyors

	A	B	C	D	E	F	G	H	I	J	K
5	R	Nu	Street	Price	Ty	Ag	Floo	Are	Rec	B	Gara
7	2	48	MouseWalk	35,000	4	60	2	545	1	1	1
11	6	32	MouseWalk	39,000	4	80	2	451	1	2	1
22	17	32	MouseWalk	47,950	4	85	2	551	2	2	2
31	26	45	MouseWalk	55,000	4	80	2	520	1	2	2
36	31	45	MouseWalk	58,950	2	60	2	640	1	3	0
38	33	57	MouseWalk	59,950	1	0	2	521	2	2	0
46	41	18	MouseWalk	64,950	3	0	2	451	1	1	1
55	50	46	MouseWalk	67,500	4	30	2	646	1	2	2
[illegible]	[illegible]	6	MouseWalk	75,000	[illegible]	55	[illegible]	650	[illegible]	[illegible]	1

(e) To turn off the Autofilter, select it again with **Menu** ⇒ **Data** ⇒ **Filter** ⇒ **Autofilter**, when all records will re-appear.

Similarly, we could select all properties with three bedrooms or all properties with two garages, or by repetition all properties with three bedrooms and two garages.

12.4 Advanced filter

This facility allows data to be filtered applying more conditions. We will use the house data example again.
We proceed as follows :

(a) Copy the data header row from row 5 and paste to row 1
(b) In row 2 we enter the criteria which we wish to apply. We will filter the list for all properties in Badger Road with prices exceeding £50,000. In row 2 under **Street** enter Badger Road, and under **Price** enter >50000.

Excel for Surveyors

	A	B	C	D	E	F	G	H	I	J	K
1	Ref	Num	Street	Price	Type	Age	Floors	Area	Recep	Bed	Garage
2			BadgerRoad	>50000							
3											
4											
5	Ref	Num	Street	Price	Type	Age	Floors	Area	Recep	Bed	Garage
6	1	70	Badger Road	35,000	2	16	2	566	2	3	0
7	2	48	MouseWalk	35,000	4	60	2	545	1	1	1

(c) Select A1:K2 and use **Menu** ⇒ **Insert** ⇒ **Name** ⇒ **Define** to name this selection "criteria"
(d) Click on **Data** ⇒ **Filter** ⇒ **Advanced Filter**. The dialogue box will ask where you would like the filtered list to appear. Click on In place and OK.

The screen should now show only those properties in Badger Road with a price exceeding £50,000.

Excel for Surveyors

	A	B	C	D	E	F	G	H	I	J	K
1	Ref	Num	Street	Price	Type	Age	Floors	Area	Recep	Bed	Garage
2			BadgerRoad	>50000							
3											
4											
5	Ref	Num	Street	Price	Type	Age	Floors	Area	Recep	Bed	Garage
26	21	1	Badger Road	53,995	2	2	2	337	1	1	0
37	32	22	Badger Road	59,950	1	0	2	462	2	2	0
41	36	99	Badger Road	61,950	3	5	2	289	1	1	1
60	55	27	Badger Road	68,950	4	15	2	504	1	3	2
63	58	97	Badger Road	71,500	4	6	2	368	1	1	2
73	68	95	Badger Road	84,950	4	5	2	416	1	2	2

More criteria can be added, but be aware that if you add too many there may be no houses in the selection. To select two streets, say Badger Road and MouseWalk, insert another row and enter MouseWalk below Badger Road in the next row. You must amend the name of the criteria range to include the additional row (A1:K3).

Excel for Surveyors

	A	B	C	D	E	F	G	H	I	J	K
1	**Ref**	**Num**	**Street**	**Price**	**Type**	**Age**	**Floors**	**Area**	**Recep**	**Bed**	**Garage**
2			BadgerRoad	>50000							
3			Mousewalk	>50000							
4											
5	**Ref**	**Num**	**Street**	**Price**	**Type**	**Age**	**Floors**	**Area**	**Recep**	**Bed**	**Garage**
6	1	70	Badger Road	35,000	2	16	2	566	2	3	0
7	2	48	MouseWalk	35,000	4	60	2	545	1	1	1
8	3	88	[illegible]	[illegible]	3	[illegible]	2	[illegible]	2	3	1

Proceed as before to filter the list.

Chapter 13

Viewing the worksheet

A disadvantage of large worksheets is that there can be too many columns to view at once. As we pointed out earlier, one sheet can have up to 256 columns and 65,536 rows. Even though you are unlikely to use all of these, Excel provides several methods to assist with the problem of extensive worksheets.

13.1 Split screen

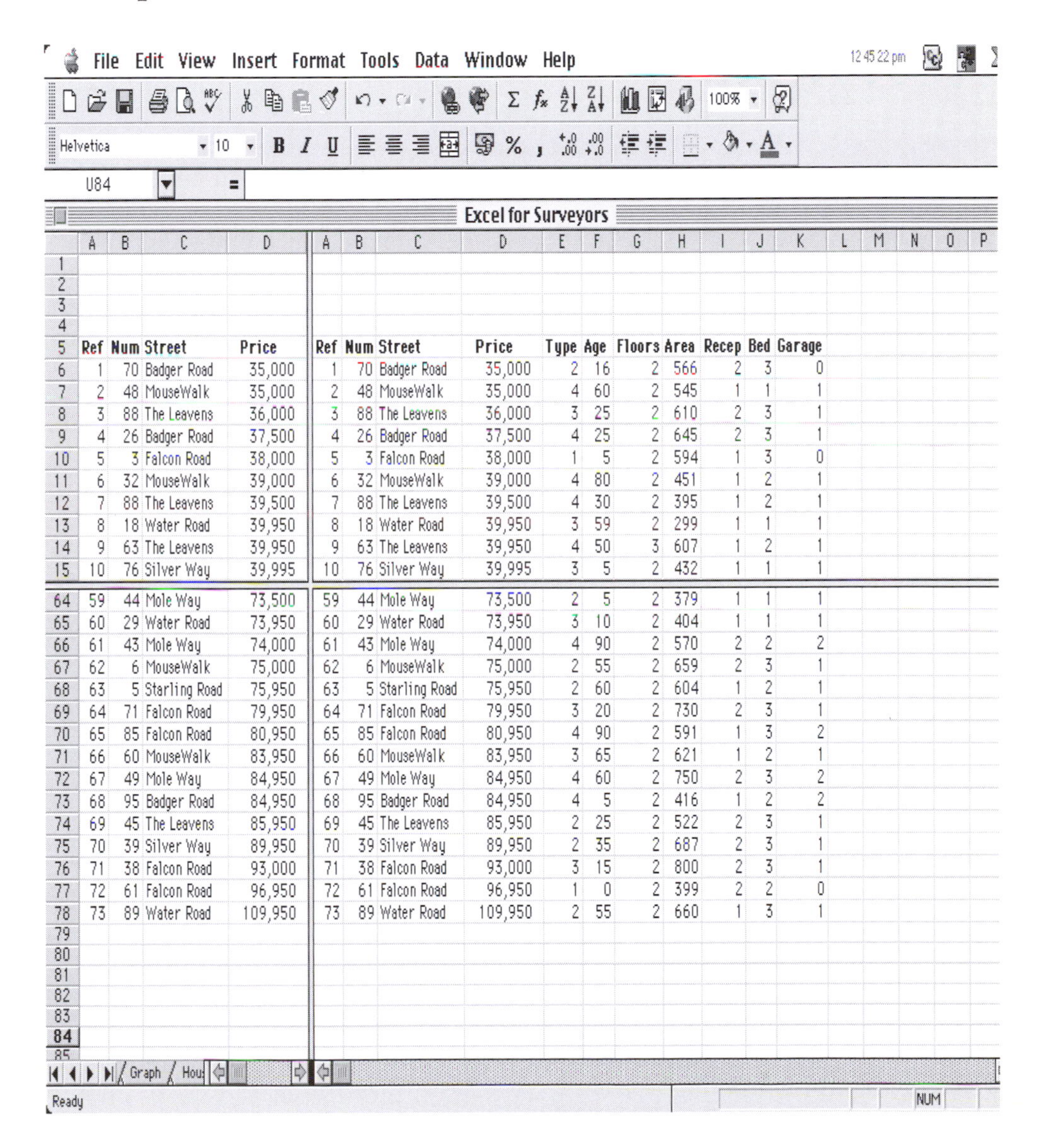

	Ref	Num	Street	Price	Type	Age	Floors	Area	Recep	Bed	Garage
5	Ref	Num	Street	Price	Type	Age	Floors	Area	Recep	Bed	Garage
6	1	70	Badger Road	35,000	2	16	2	566	2	3	0
7	2	48	MouseWalk	35,000	4	60	2	545	1	1	1
8	3	88	The Leavens	36,000	3	25	2	610	2	3	1
9	4	26	Badger Road	37,500	4	25	2	645	2	3	1
10	5	3	Falcon Road	38,000	1	5	2	594	1	3	0
11	6	32	MouseWalk	39,000	4	80	2	451	1	2	1
12	7	88	The Leavens	39,500	4	30	2	395	1	2	1
13	8	18	Water Road	39,950	3	59	2	299	1	1	1
14	9	63	The Leavens	39,950	4	50	3	607	1	2	1
15	10	76	Silver Way	39,995	3	5	2	432	1	1	1
64	59	44	Mole Way	73,500	2	5	2	379	1	1	1
65	60	29	Water Road	73,950	3	10	2	404	1	1	1
66	61	43	Mole Way	74,000	4	90	2	570	2	2	2
67	62	6	MouseWalk	75,000	2	55	2	659	2	3	1
68	63	5	Starling Road	75,950	2	60	2	604	1	2	1
69	64	71	Falcon Road	79,950	3	20	2	730	2	3	1
70	65	85	Falcon Road	80,950	4	90	2	591	1	3	2
71	66	60	MouseWalk	83,950	3	65	2	621	1	2	1
72	67	49	Mole Way	84,950	4	60	2	750	2	3	2
73	68	95	Badger Road	84,950	4	5	2	416	1	2	2
74	69	45	The Leavens	85,950	2	25	2	522	2	3	1
75	70	39	Silver Way	89,950	2	35	2	687	2	3	1
76	71	38	Falcon Road	93,000	3	15	2	800	2	3	1
77	72	61	Falcon Road	96,950	1	0	2	399	2	2	0
78	73	89	Water Road	109,950	2	55	2	660	1	3	1

At the top of the vertical scroll bar is a small solid grey (PC) or black (Macintosh) rectangle. To split the worksheet drag this down and the screen will be divided horizontally into two sections by two parallel lines. Each section of the worksheet will now have its own scroll bar and this can be dragged to view any part of the worksheet without affecting the other section. Similarly there is a rectangle on the right of the horizontal scroll bar which can be dragged to split the screen vertically. Of course, if you drag both rectangles you can divide the screen into four sections which can be scrolled independently. However the worksheet is still a single entity and operations are not affected at all.

To move to another section click any cell in that section. To cancel the split screen just drag the black rectangles back to their normal position, but note that your view of the screen will be that in which the active cell is.

(If you are using Freeze panes, described below, you cannot apply split screen.)

13.2 New window

A similar method of seeing two or more parts of a worksheet at once is to use the **Menu ⇒ Window ⇒ New Window** command. This creates an independent window into the worksheet with its own scroll bars. However, as with split screens, the worksheet is still an entity and alterations made in one window will be seen in the other window if the cell is visible there. Each window can be split into two or four panes, and several windows can be created, so that a proliferation of views can be seen at once. However, taken to extremes this can create greater confusion rather than less.

An advantage of a new window as compared with a split screen is that you can select another worksheet in the same workbook and see both at the same time.

13.3 Freeze panes

If the top row and/or the left-hand column of a large data set contain headings for the data we may wish to be able to see these when looking at data in distant parts of the worksheet. Freezing panes is a very convenient way to do this. To freeze a row or column or both, select the cell which is immediately below and to the right of the row and column to be frozen and select **Menu ⇒ Window ⇒ Freeze panes**. Black lines will appear dividing the panes and it will then be possible to scroll through the worksheet. As you scroll around the row and column headings at the top and left of the sheet will move to match so that it is always possible to see the headers for the selected cell.

The house data from our previous example would appear as opposite. Rows have been scrolled down to Row 14, but the headings of the columns are visible above.

(If you are using split screen, applying freeze panes will convert the screen to freeze mode at the point of the split.)

13.4 Group columns

There may be occasions when you need to preserve columns within a sheet for reference, but do not normally want to see these. They can be hidden using the group columns command.

File Edit View Insert Format Tools Data Window Help

Helvetica 10

T61 =

Excel for Surveyors

	A	B	C	D	E	F	G	H	I	J	K
1	Ref	Num	Street	Price	Type	Age	Floors	Area	Recep	Bed	Garage
22	17	32	MouseWalk	47,950	4	85	2	551	2	2	2
23	18	40	Badger Road	48,950	1	20	2	660	1	2	0
24	19	95	Starling Road	51,950	3	90	2	440	1	2	1
25	20	54	Falcon Road	53,000	3	80	2	652	2	3	1
26	21	1	Badger Road	53,995	2	2	2	337	1	1	0
27	22	44	The Leavens	54,000	3	2	2	293	1	1	1
28	23	32	Falcon Road	54,495	3	60	2	568	1	2	1
29	24	22	Starling Road	54,500	3	50	2	761	2	3	1
30	25	81	Falcon Road	54,500	4	8	2	267	1	1	2
31	26	45	MouseWalk	55,000	4	80	2	520	1	2	2
32	27	76	Starling Road	56,950	1	0	2	320	1	2	0
33	28	82	The Leavens	57,000	4	80	2	624	2	2	2
34	29	58	The Leavens	57,995	2	55	2	650	2	2	0
35	30	48	Falcon Road	58,450	2	20	2	390	1	1	0
36	31	45	MouseWalk	58,950	2	60	2	640	1	3	0
37	32	22	Badger Road	59,950	1	0	2	462	2	2	0
38	33	57	MouseWalk	59,950	1	0	2	521	2	2	0
39	34	13	Water Road	59,950	4	80	2	455	2	2	2
40	35	95	Silver Way	59,950	2	20	2	700	2	3	1
41	36	99	Badger Road	61,950	3	5	2	289	1	1	1
42	37	58	Mole Way	62,000	4	10	2	310	1	1	2
43	38	46	The Leavens	62,500	4	50	2	520	2	2	2
44	39	33	Silver Way	62,500	4	50	2	543	2	3	2
45	40	74	Mole Way	62,500	1	0	2	533	2	2	0
46	41	18	MouseWalk	64,950	3	0	2	451	1	1	1
47	42	54	Starling Road	64,950	2	40	2	588	2	2	1
48	43	31	Falcon Road	64,995	4	5	2	452	1	2	2
49	44	68	The Leavens	65,000	3	90	2	554	2	2	1
50	45	59	Mole Way	65,000	3	60	2	540	1	1	1
51	46	9	Falcon Road	65,000	4	9	2	478	1	2	2
52	47	74	Mole Way	65,000	2	10	2	730	2	3	1
53	48	85	Falcon Road	65,500	3	5	2	316	1	1	1
54	49	13	Starling Road	67,500	4	20	2	688	1	3	2
55	50	46	MouseWalk	67,500	4	30	2	646	1	2	2
56	51	22	Water Road	67,500	4	25	2	685	1	3	2

OverRent / Chart / Graph / HouseTrend / Residual / S-Curve / Profits / Filter

Ready

The example below is of the accounts for the collection of three ground rents. It contains a total of 48 columns, showing, for each lessee and for each half year, the arrears, rent, date rent due, total, sum paid and date paid. Arrears are carried forward and sums due and paid calculated each half year. For the purposes of collection it is not normally necessary to see the record of previous payments but the details must be preserved.

Excel for Surveyors

	A	B	C	D	E	F	G	H	I	J	K	L	M
1	Names	Street	District	Town	County	Postcode	Arrears pre-1997/1	Rent 1997/1	Due date 1997/1	Total 1997/1	Paid for 1997/1	Date 1997/1	Arrears pre97/2
2	Mr C Adams	1 Bournemouth Road	Somewhere	Anytown	Queshire	AB1 2YZ	25.00	25.00	1 Jan 97	50.00	25.00	3 Jan 97	25.00
3	Mr and Mrs Gould	2 Bournemouth Road	Somewhere	Anytown	Queshire	AB1 2YZ	0.00	25.00	1 Jan 97	25.00	25.00	1 Feb 97	0.00
4	Mr Robertson	3 Bournemouth Road	Somewhere	Anytown	Queshire	AB1 2YZ	50.00	25.00	1 Jan 97	75.00	50.00	20 Jan 97	25.00
5													

To do this select the columns to be hidden and then **Menu ⇒ Data ⇒ Group and Outline. . . ⇒ Group**. A bar will appear across the top of the selected cells and on the right hand end there will be a box containing a "–" sign. Click on this and all the selected columns will disappear, leaving only, in this case, the entries for the current half year, and a "+" sign above the next column.

Excel for Surveyors

	A	B	C	D	E	F	AQ	AR	AS	AT	AU
1	Names	Street	District	Town	County	Postcode	Arrears pre-00/1	Rent 2000/1	Due date 2000/1	Total 2000/1	Paid for 2000/1
2	Mr C Adams	1 Bournemouth Road	Somewhere	Anytown	Queshire	AB1 2YZ	25.00	25.00	1 Jan 00	50.00	
3	Mr and Mrs Gould	2 Bournemouth Road	Somewhere	Anytown	Queshire	AB1 2YZ	0.00	25.00	1 Jan 00	25.00	
4	Mr Robertson	3 Bournemouth Road	Somewhere	Anytown	Queshire	AB1 2YZ	25.00	25.00	1 Jan 00	50.00	

All columns are still there and can be seen by clicking on the "+". Of course cells which would be altered by change in a cell elsewhere will still be altered while they are hidden.

13.5 Forms

If we wish to see just the data relating to a particular record (row) we can use the **Forms** command. However this has the restriction that only 32 columns can be shown.

As an example we take the accounts worksheet, and proceed :

Since there are 48 columns in the full worksheet, we group and hide columns G to X which relate to previous years.

Select rows 1–4 which include the column headings and all the data rows.

Select **Menu ⇒ Data ⇒ Form . . .** and a panel will appear showing all the data relating to the first entry.

Each value for the first row appears in a box to the right of the corresponding field name (column heading) and can be altered or amended within the box. Note that some of the values are in the grey area and not within a box – these are computed values and cannot be altered. When rent due on 1 January 2000 is received this can be entered in the appropriate box and will be automatically copied to the worksheet.

By dragging the scroll bar down or clicking on the "Find next" or "Find prev" buttons the other records can be seen.

We can also select a subset of records by using the "Criteria " button. Using our previous example of the House Data select **Menu ⇒ Data ⇒ Record** and then **Criteria**. Enter "MouseWalk" into the "Street" box and click on **Form**. The **Find Next** button will then look for the next occurrence of that street.

There are other uses for forms, notably the ability for new records to be input by other persons or perhaps copied from printed forms issued as part of a survey, but these will not be considered here.

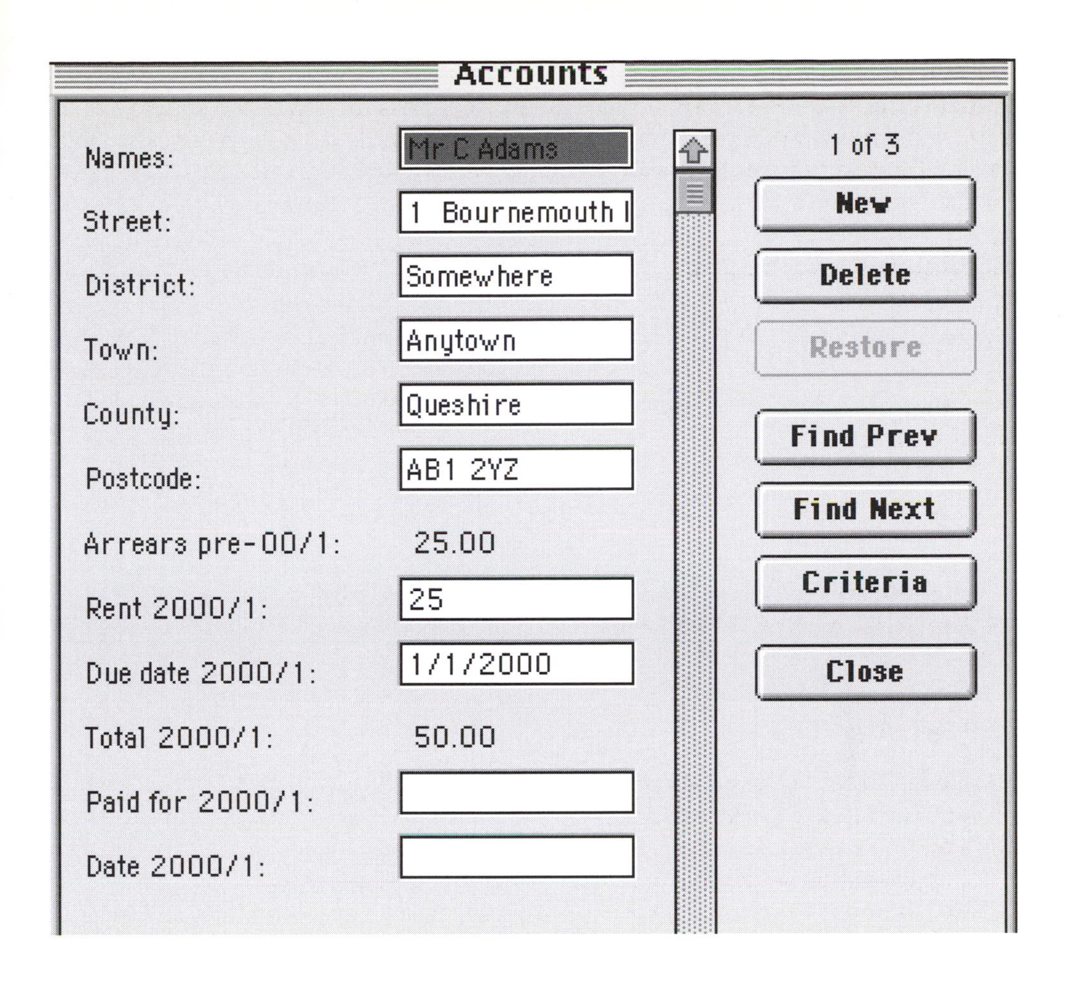
Accounts
Names:
Mr C Adams
Street:
1 Bournemouth l
District:
Somewhere
Town:
Anytown
County:
Queshire
Postcode:
AB1 2YZ
Arrears pre-00/1:
25.00
Rent 2000/1:
25
Due date 2000/1:
1/1/2000
Total 2000/1:
50.00
Paid for 2000/1:
Date 2000/1:
1 of 3
New
Delete
Restore
Find Prev
Find Next
Criteria
Close

Chapter 14

More uses for Excel

14.1 Pivot tables

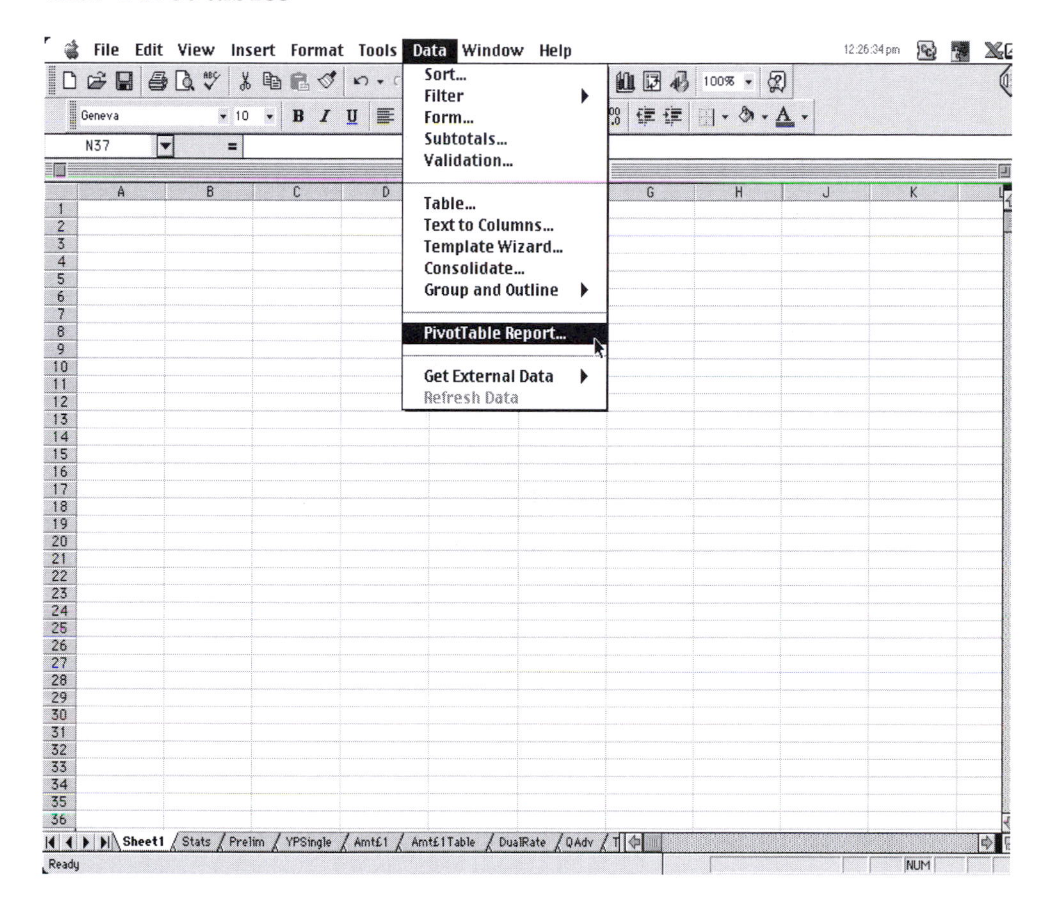

As we have seen, an Excel worksheet can hold a very large amount of data. Summarising data in such a large sheet can be simplified with the use of Pivot Tables, which can be set up with the Pivot Table Wizard.

Example

For this example we use the same data as we considered in the case of sorting and filtering. We wish to summarise the data from chapter thirteen by road and by type, finding the total price obtained for each class.

The steps are :

(a) Select **Menu** ⇒ **Data** ⇒ **Pivot Table** . . .
(b) Select **Microsoft Excel List** or **Database**.

(c) Select the data by dragging over the whole area of the cells, including the titles in the top row.

(d) Drag the titles of the various columns into the Pivot Table boxes. In this case drag Street into Row, Type into Column and Price into Data. (Note that some of the variables cannot be seen at first, but will come into view if you move the scroll bar.)

(e) Select starting cell. In this case select L4. If you do not make a selection Excel will create a new worksheet for the output.

(f) Click **Finish**. Excel will now create a table of totals of values for each street and for each type, together with the grand total.

L	M	N	O	P	Q
Sum of Price	Type				
Street	1	2	3	4	Grand Total
Badger Road	108900	88995	61950	262900	522745
Falcon Road	134950	102400	345945	309395	892690
Mole Way	62500	208450	65000	220950	556900
MouseWalk	59950	133950	148900	244450	587250
Silver Way		149900	39995	62500	252395
Starling Road	125900	208400	106450	67500	508250
The Leavens		188895	155000	312400	656295
Water Road		109950	159400	238450	507800
Grand Total	492200	1190940	1082640	1718545	4484325

It may be however that we require other information, for example the number of houses rather than the total value. To change to this:

(g) select a cell within the table and click **Menu** ⇒ **Data** ⇒ **Pivot Table Report**

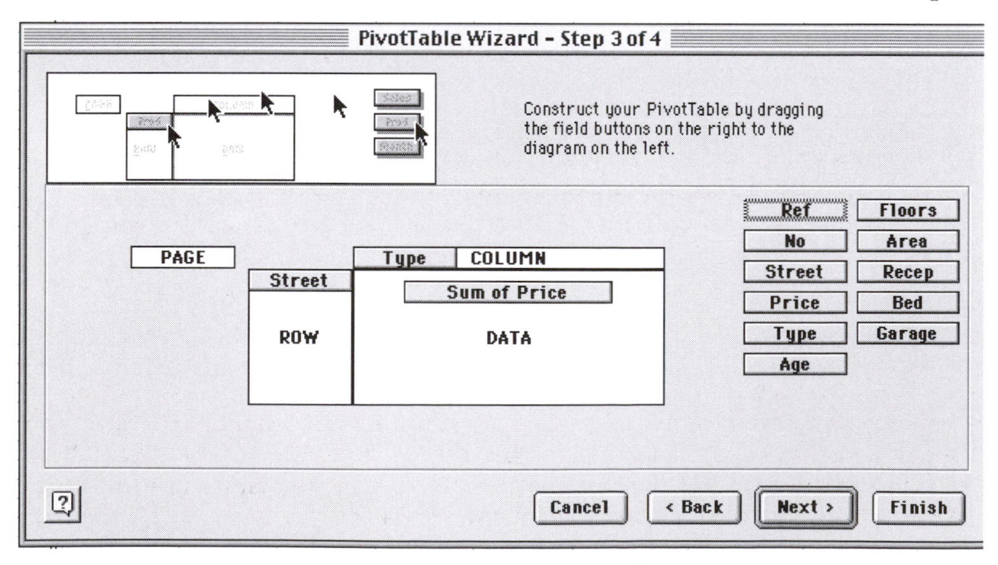

(h) in the **Data** area double click on "Sum of Price" and the Pivot Table Field dialogue box will appear.

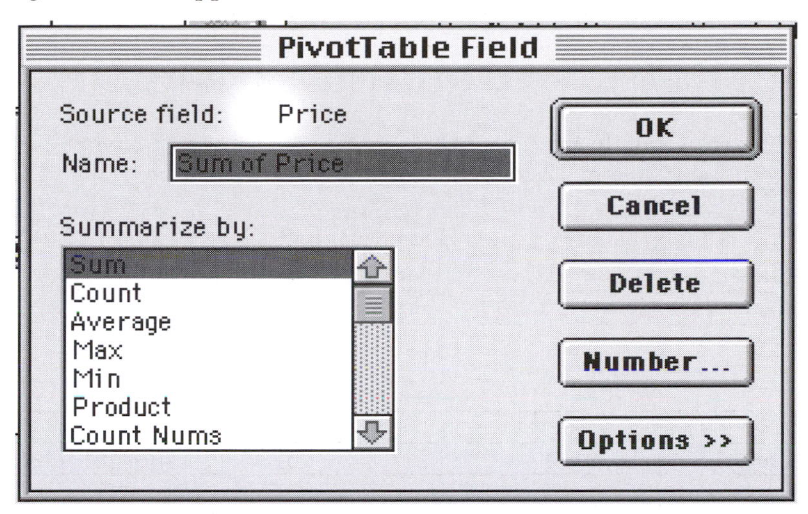

(i) Click on **Count** and **Finish** and the Pivot Table will show the number of houses in each category.

Count of Price	Type				
Street	1	2	3	4	Grand Total
Badger Road	2	2	1	4	9
Falcon Road	2	2	5	5	14
Mole Way	1	3	1	3	8
MouseWalk	1	2	2	5	10
Silver Way		2	1	1	4
Starling Road	2	3	2	1	8
The Leavens		3	3	6	12
Water Road		1	3	4	8
Grand Total	8	18	18	29	73

You will see that there are several other options – we leave you to experiment with these.

14.2 Data tables

We have already seen that it is possible to do "What if" calculations very easily for example in the case of term and reversion valuations. One can experiment with changes in the rate of interest, reversionary rent, etc. to test the resulting valuation. The **Goal Seek** command which we examined is a very useful and convenient method for adjusting one variable to give the target result.

But what if we wish to see the effect of several combinations of two different variables on the result? Using the fully let property at the beginning of chapter 7, we might be interested in looking at the effect of a combination of different rents and comparable yields. A sensitivity analysis no less. Instead of plugging in the changes to our fully let calculation we can use a data table.

Before we initiate this function we need to set up a matrix. This will state the value we want to find for each combination of variables, and then the combination of variables themselves.

In the following example, which can be placed anywhere on the worksheet, the value we want to find (in this case the fully let valuation) is placed as the top left of our matrix. This is done by referencing the cell to the original answer/valuation. Our choice of variations in rent have been input as the row across the top of the matrix and the variations in yield have been input as the column to the left of the matrix. In this example :

Cell C6 is set equal to C4;
Cells D6:G6 are the rates of interest we wish to consider;
Cells C7:C10 are the possible rents.
Cells D7:G10 – those in the empty box – are where the answers will appear. (The box is drawn only for ease of identification).

Excel for Surveyors

	A	B	C	D	E	F	G
1	FULLY LET PROPERTY						
2	Rent		8,000				
3	YP Perp	8%	12.5				
4			100,000				
5							
6			100,000	6.50%	7%	7.50%	8%
7			7,000				
8			7,500				
9			8,000				
10			8,500				
11							

Now that we have set up the matrix we want to fill it with the different valuations that result from the respective combination of variables. To do this we initiate the Data Table command.

(a) Highlight the whole matrix, from the valuation to the last empty cell – C6:G10
(b) Click on **Menu** ⇒ **Data** ⇒ **Table**.
(c) The dialog box asks two questions:
   (i) what is the row input cell? Enter the absolute reference to the original yield in the original calculation, (B3); and
   (ii) what is the column input cell? Enter the absolute reference to the original rent (C2).
(d) Click OK and your matrix should fill with all the different valuations.

	A	B	C	D	E	F	G
1	FULLY LET PROPERTY						
2	Rent		8,000				
3	YP Perp	8%	12.5				
4			100,000				
5							
6			100,000	6.50%	7%	7.50%	8%
7			7000	107,692	100,000	93,333	87,500
8			7500	115,385	107,143	100,000	93,750
9			8000	123,077	114,286	106,667	100,000
10			8500	130,769	121,429	113,333	106,250
11							

You may find that format of the numbers are not what you expected. If this is the case simply highlight the valuations in the matrix and change them to a more suitable format by clicking on one of the number format buttons, on the formatting tool-bar.

Finally, you could summarise these results perhaps by finding the mean of these valuations with the **Menu ⇒ Insert ⇒ Function ⇒ Statistical ⇒ Average** or **Menu ⇒ Insert ⇒ Function ⇒ Statistical ⇒ Median** commands as discussed in chapter 5.

14.3 Scenarios

A common method of approaching a problem where inputs are uncertain is the use of scenarios. Very often three scenarios are tried – best case, worst case, and most likely outcome. Noting the results of "What if" calculations on paper (or printing) can enable us to study the variations, but Excel provides a method for organising results on screen.

We assume for the purposes of this example the valuation shown below which we consider to be the most likely. However we wish to look at best-case and worst-case scenarios. We will assume the following:

	Most likely outcome	Worst outcome	Best outcome
Rent on reversion	£7,000	£6,000	£7,500
Repairs on reversion	£500	£600	£450
Yield on reversion	8%	8.5%	7.75%

We first set up the worksheet in the usual way for the most likely outcome. Having set up the worksheet, we then

(a) Select **Menu ⇒ Tools ⇒ Scenarios ⇒ New** and the Scenario Manager will appear.
(b) Select Add and enter a name for a scenario – "Worst".

	A	B	C	D
1				
2				
3	Rent		5,000	
4	less external repairs		500	
5			4,500	
6	YP 4 years	7.%	3.3872	15,242
7	Reversion to		**7,000**	
8	less external repairs		**500**	
9			6,500	
10	YP perp deferred 4 years	**8.%**	9.1879	59,721
11				74,964
12				

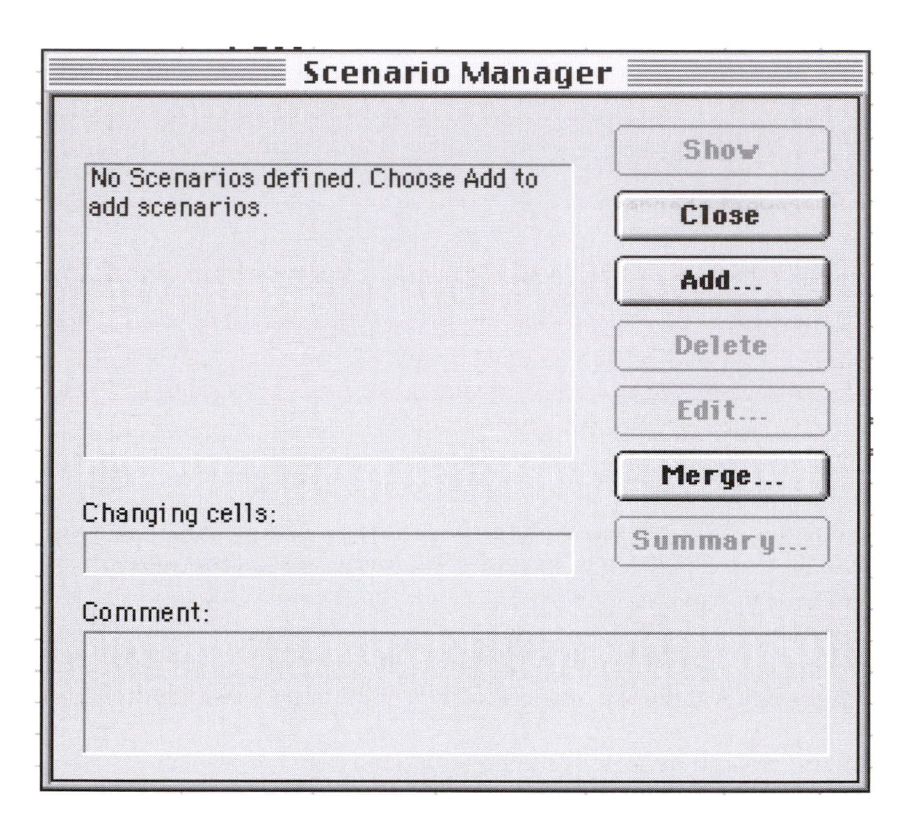

(c) Enter the cells which we want to change in our "worst case" scenario by clicking on each of them – C7, C8 and B10. Hold down the Control[25] key while you do this if they are not adjacent.

[25] On the Macintosh use the Command key.

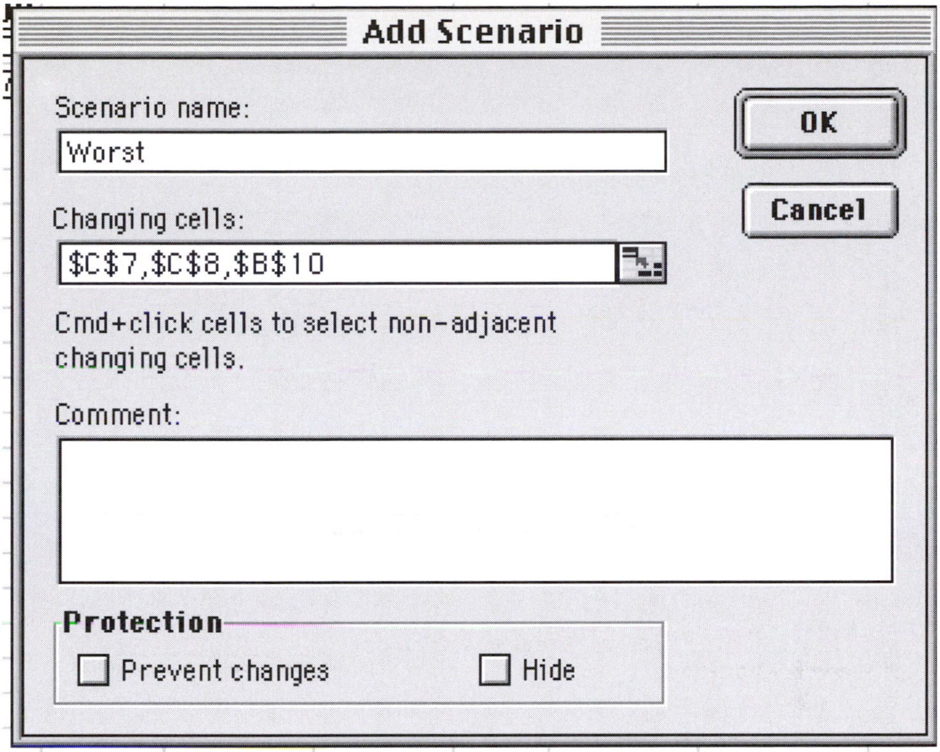

Add Scenario

Scenario name:

Worst

Changing cells:

C7,C8,B10

Cmd+click cells to select non-adjacent
changing cells.

Comment:

Protection

Prevent changes

Hide

OK

Cancel

(d) Add a comment if you wish. Otherwise the name of the person in whose name this copy of Excel is registered will be entered by default (see below).

(e) Click OK and you will be prompted for any changing data in these cells. Change C7 to 6000, C8 to 600 and B10 to 8.5% and then click OK to accept this new data.

(f) Click OK and you will return to the Scenario manager.

Repeat the process to add other scenarios, perhaps a "Best Case" scenario.

When you have set up all the scenarios you wish, click on **Summary** . . . in the Scenario Manager and the scenarios will be calculated and placed in a new Worksheet alongside the one you have been using to create it.

This worksheet can be adjusted in the usual way, but you can also see some small + and – boxes at the top and left. These enable you to see or hide parts of the scenario.

More scenarios can be set up based on the same data.

14.4 Validation of data

When entering data we may wish to ensure that a particular cell or group of cells will only accept data of a particular type, for example, a date. This can be done with the **Menu** ⇒ **Data** ⇒ **Validation** command. Select the cell which you wish to restrict with this command and click on the allow arrow, This will give a series of options, select **Date**. You then have the option to select some more restrictions – select **Greater than or equal to** and enter 1/1/99 in the start date box and click

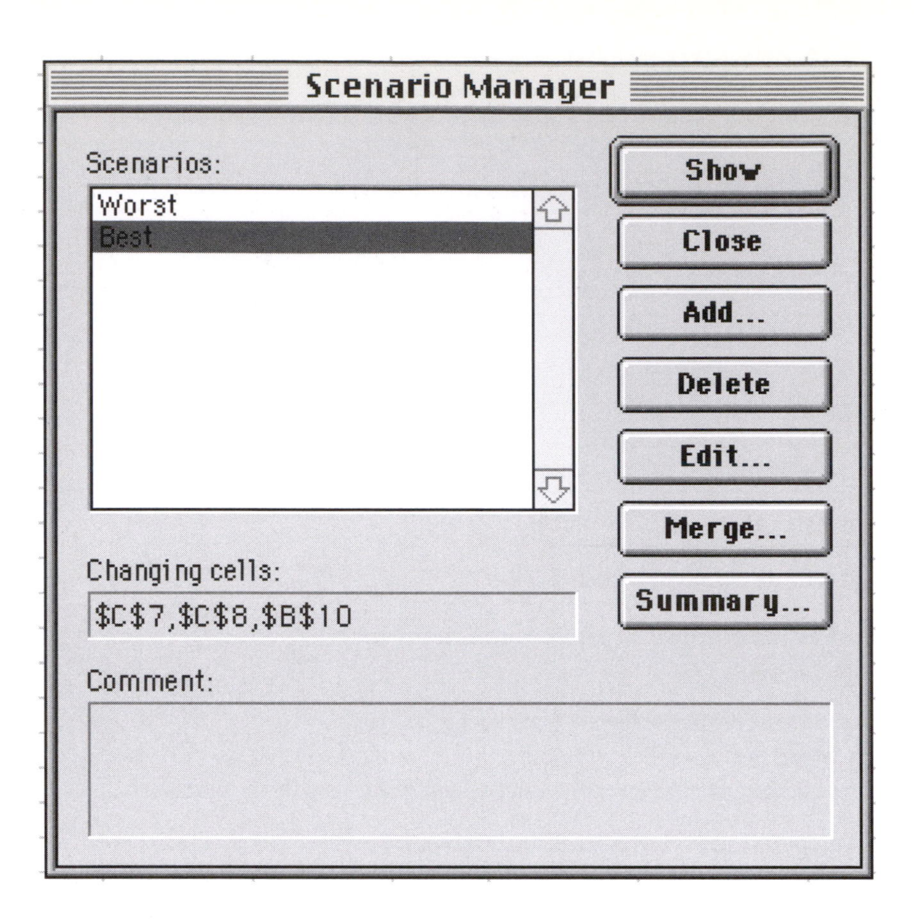

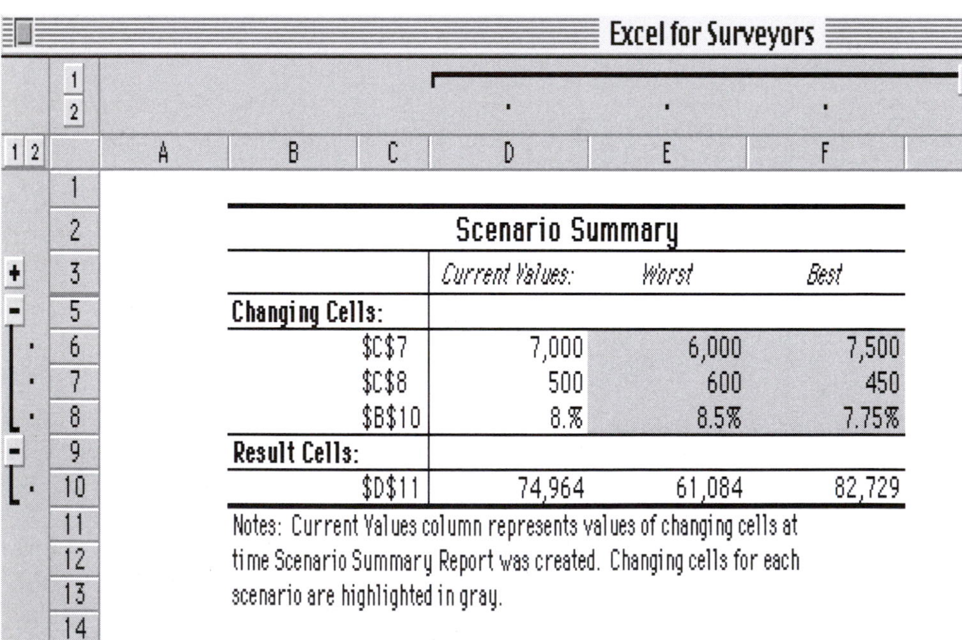

Scenario Summary

	Current Values:	*Worst*	*Best*
Changing Cells:			
C7	7,000	6,000	7,500
C8	500	600	450
B10	8.%	8.5%	7.75%
Result Cells:			
D11	74,964	61,084	82,729

Notes: Current Values column represents values of changing cells at time Scenario Summary Report was created. Changing cells for each scenario are highlighted in gray.

OK. You will now not be able to enter anything in the cell except a date in one of the recognised formats on or after 1 January 1999. The format of the date can of course be changed as described above.

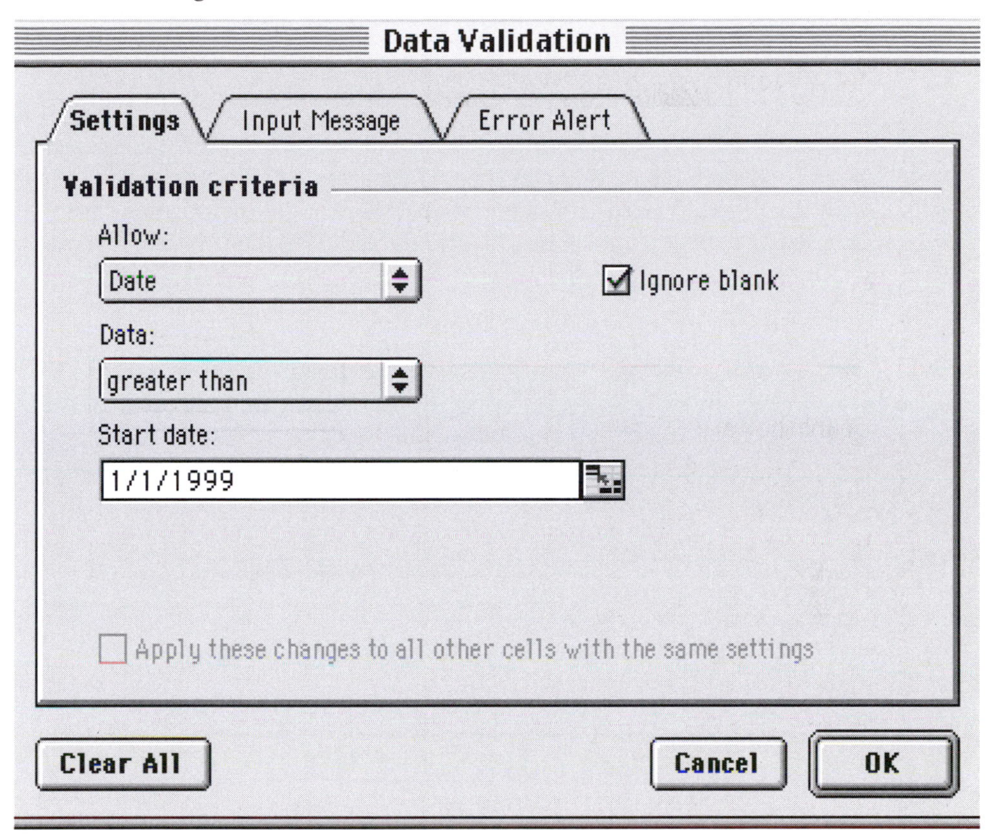

Click on the allow arrow in data validation again and this time select list. This is useful if you want to allow only the available tax rates in a particular cell, or perhaps the rates of personnel in the company. The next text box will ask you to indicate source. This allows you to select cells on an existing or another spreadsheet that contains the entries that you would like to restrict the cell to. However you may just prefer to enter the information (which can be text or numbers) straight into the text box. To do this separate each possible entry with a comma. For different hire rates you would enter say:

£60,£75,£90,£ 120

Click OK and a little down arrow will appear to the right of the cell. By clicking on this the list will appear. Select an item from the list and it will stay in the cell. The down arrow will only be apparent if the cell is selected.
If you enter any number other than those in the list an error message will appear. It is in fact possible to specify your own error message. Click on the "Error Alert" tab in **Data Validation**. On the left hand side, it is possible to choose between stop, warning and information signals. On the right hand side are text boxes available

for your own title and message. Apart from the many possible humorous uses, this is useful for telling a third party user or reminding yourself why you cannot enter a particular number into the cell. You can also create your own prompt. This is the **Input Message** tab. We leave you to experiment with. these.

14.5 A simple risk model using the SUMPRODUCT function

The scenario summary is a quick and easy way to obtain a risk adjusted measure. Having created the most likely, worst and best scenarios and found the respective effects on the result a naive risk adjusted return would be to average the results. This simply involves adding up the results and dividing by the number of results,

$$\text{ie} \quad (74{,}964 + 61{,}084 + 82{,}729) \;/\; 3$$
$$= 72{,}926$$

This is a naive average. In other words, each of the scenarios are given a 33.33% chance of occurring. Instead we could apply our own probabilities to each of the scenarios. So for example most likely would mean exactly that and we could apply a 50% probability. We could apply 20% to the worst case and 30% to the best case. You could put these probabilities in the top row as follows.

Excel for Surveyors

Scenario Summary

	Current Values:	*Worst*	*Best*
Changing Cells:	50%	20%	30%
C7	7,000	6,000	7,500
C8	500	600	450
B10	8.%	8.5%	7.75%
Result Cells:			
D11	74,964	61,084	82,729

Notes: Current Values column represents values of changing cells at time Scenario Summary Report was created. Changing cells for each scenario are highlighted in gray.

Risk adjusted value £74,517

Note the probabilities MUST add up to 100%. To obtain this new risk adjusted measure, multiply the 50% by the most likely result, the 20% by the worst result, and 30% by the best case result. Finally add them together.

Depending on the number of scenarios this could be an increasingly laborious calculation in Excel. However there is a function that will make this a much simpler operation. The SUMPRODUCT function takes two lists of numbers as arguments. Given these, it will multiply the corresponding pair from each list and then total them.

So instead of

= (D5 * D10) + (E5 * E10) + (F5 * F10)

select the SUMPRODUCT function and the two sets of numbers just as you would the one set of numbers in the sum function.

SUMPRODUCT(D5:F5,D10:F10)
= 74,517

A note of caution: Probabilities are very debatable and as such should be carefully applied. Thorough explanation should always be attached to such assumptions.

Chapter 15

Other facilities in Excel

15.1 Macros and Visual Basic

A macro is simply a recording of a series of steps which can then be repeated by clicking on a custom-created button. These can perform elementary operations such as formatting a cell in a predefined way to avoid repetitive instructions which would take much longer to perform by the operations already described.

One of the first "high-level" computing languages used in small computers was Basic, and various versions of this were built into IBM and Apple machines in the late 1970s. Very few commercial programs were available at that time, and one often had to sit down and write a program for a particular purpose before actually using it.

The original incarnations of Basic eventually became too slow and inefficient for modern operations and are hardly ever used these days. However, it is sometimes convenient to give instructions to Excel to perform operations which would otherwise require considerable manipulation. Such operations can be created as a macro which is written in a form of Basic language within Excel. We discussed the construction of standard valuation tables earlier, but it may be of interest to note that such tables could be created on the click of a single button programmed to run an appropriate macro.

Some knowledge of the original Basic is of considerable help in constructing and editing an Excel macro, and while the initial operation of recording is relatively straightforward the main advantage of Basic is the ability to set up repetitive routines with loops.

The operation is discussed in the manuals which accompany your software, but you may consider it preferable to leave a study of these until you have mastered the principles discussed earlier.

15.2 Monte Carlo simulation

This technique is an extension of the Scenario approach. It is clear that in many situations some of the input variables could take on several different values, for example, should the value on reversion be £150 per m^2, £175, or some number in between. Allowing for all the possible values of each variable and recalculating for each combination, the total number of possibilities could easily run into millions, giving a range of valuations from which the mean and standard deviation could be calculated.

Fortunately a good approximation to these statistics can be obtained by doing just a few hundred of the total possibilities provided we arrange for each of the inputs to vary randomly over a range which we define for each valuation. These results can be then entered on a chart from which the general pattern can be visualised in addition to any numerically expressed statistics.

We do not include the procedure for simulation here, but it may be relevant to note that Visual Basic can be used to set up such a procedure, when 100 or 1,000 valuations can be done in a few seconds and statistics and charts produced immediately.

15.3 Regression analysis

Regression analysis is a technique used in many different disciplines for estimating the significance of a number of factors (the independent variables) against another factor (the dependent variable).

For example, it is possible to estimate the relative significance of the factors which contribute to the value of dwelling-houses and obtain coefficients for each independent variable. We can then make an estimate of the value of similar houses. There will inevitably be a margin of error in such an estimate, but other statistics can give a measure of this.

Excel will carry out regression analysis on data, provided that there is sufficient to be significant. As a rule of thumb, one should have at least 10 observations for each independent variable, so that for example, if we are seeking the 10 most significant factors in determining the price of houses we should need at least 100 cases. These should not include "outliers" – unusual cases not regarded as typical, such as very large detached houses or those with an attractive sea view.

The procedure for doing regression analysis is found in the Add-ins, if they have been installed, under **Menu** ⇒ **Tools** ⇒ **Data Analysis** . . . ⇒ **Regression Analysis**, and some explanation of the operation will be found in the Help Menu.

15.4 Programming[27]

Paradoxically this is not about writing computer programs. It is about situations where we seek to optimise a solution which is subject to one or more constraints, and mathematical programming is concerned with this. Some problems in valuation require a more complicated approach than that which Goal Seek can provide and the **Menu** ⇒ **Tools** ⇒ **Solver** command offers a method for doing this.

---

[27] A detailed and fairly readable explanation of programming may be found in *Encyclopaedia Britannica* under "The Mathematical Theory of Optimisation" (though there is some heavier maths stuff towards the end).

Chapter 16

Printing

File Edit View Insert Format Tools Data Window Help

New... ⌘N
Open... ⌘O
Close ⌘W
Save ⌘S
Save As...
Save as HTML...
Save Workspace...
Page Setup...
Print Area
Print Preview
Print... ⌘P
Properties...
1 Data:BAYFIELD:Excel for Surveyors
Quit ⌘Q

Excel for Surveyors

					E	F	G	H
5				58,500				
6				00,000				
7				58,500				
8				1.2857				
9					9,407,143			
12				75,000				
13				50,000				
14				25,000				
16				,000				
17				4,125,000				
18	Professional fees	12.5%		515,625				
19	*Short-term finance*			4,640,625				
20	Contingencies @	3.0%		139,219				
21				4,779,844				
22	building costs etc	14.0%		493,576				
23				5,273,420				
24	for 3 months'	14.0%		357,052				
25				5,630,471				
26	Letting and Sale	15.0%	98,775					
27	Advertising and		25,000					
28	Fees for selling	2.0%	188,143	311,918				
29				5,942,389				
30	Return for risk and	17.0%		1,599,214				
31	TOTAL EXPECTED				7,541,604			
32	Site value on				1,865,539			
33	PV £1	14.0%	2.5 years		0.72067			
34					1,344,443			
35	Less acquisition	2.5%			33,611			
36	TODAY				1,310,832			

Graph / HouseTrend / Residual / S-Curve / Profits / Filter / Pivot / Scenario Su

Ready

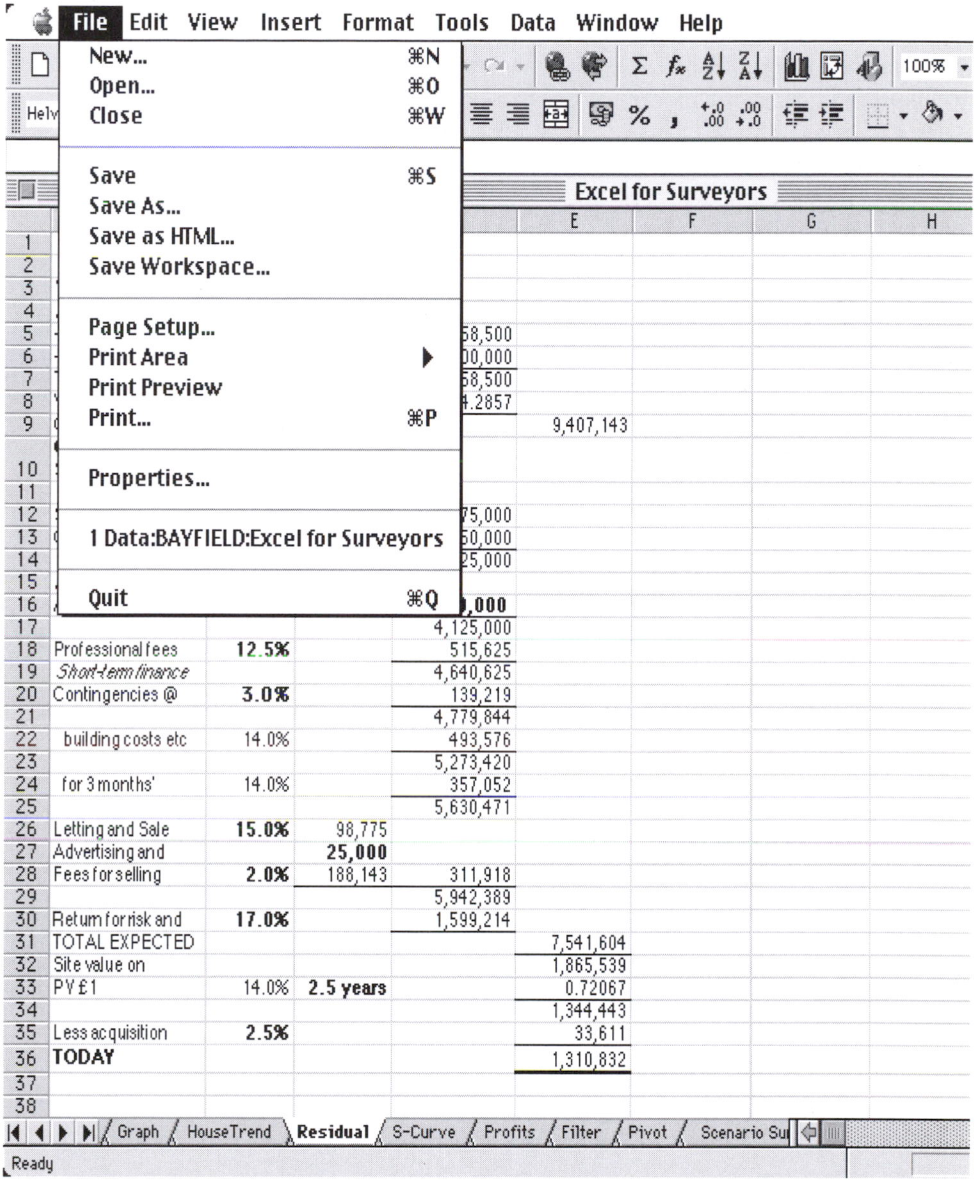

16.1 Making your work accessible

Once you have set up and analysed your data in Excel you will probably want to circulate it and show it to other people: your manager, your staff or your clients maybe? There are several ways in which you can do this: printing; saving to

floppy disk; transferring the data to a Word file or a Power-Point presentation, e-mail, or publishing on the web.
We will concentrate first on printing data directly from Excel.

16.2 Page set-up

Click on **Menu** ⇒ **File** ⇒ **Page Set-up** to obtain a dialog box which pops up with various options. Normally the contents of the first and most commonly used tab, **Page** will be visible. If it is not then select the **Page** tab.

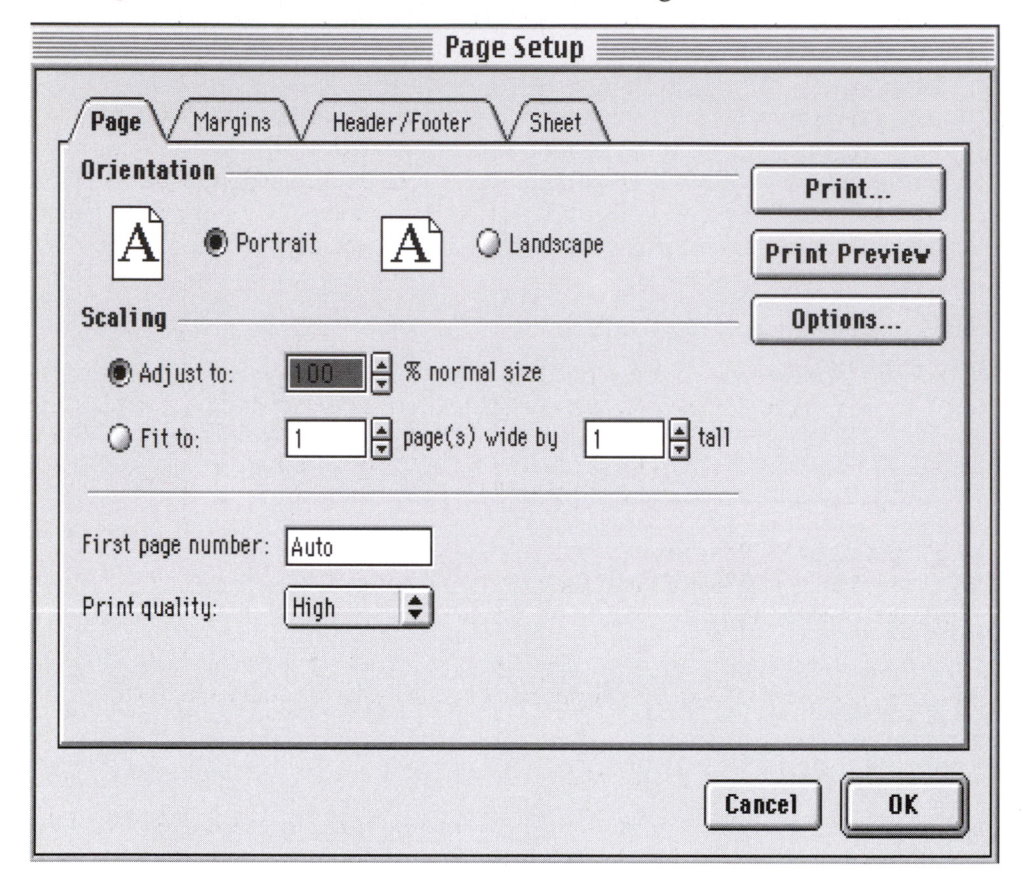

16.2.1 The Page Tab

Under the Page tab are several important options to control your printouts.

(a) Orientation and scaling. By default Excel orientates your work to a portrait format (taller). Select landscape (wider) if you want to turn the page on its side. This is invaluable when printing cash flows and large tables, and you want to fit as much as possible of your work onto one page.

(b) There are two options to allow you to scale the size of your printing

  (i) Reduce (or increase) by a chosen percentage;

  (ii) Use the **Fit To** button and enter the number of pages to which your work is to be fitted. The default is to fit to one page but you can make it fit to as many or as few pages as you care to select. When you select the **Fit to** option button the % normal size text box above automatically changes to tell you by how much your work has been decreased or increased in size. Should you wish to specify your own size you can alter the percentage in the text box yourself.

(c) First Page No. The default is auto which normally starts numbering at page 1. You can alter this to any number you wish. However page numbers will not actually appear unless inserted in the header or footer which we discuss below.

(d) Print quality. You can vary the quality by selecting a higher or lower "dpi" (dots per inch). The more dots the higher the quality, but the longer it will take to print.

(e) An options button gives you a further opportunity to change the scaling and portrait/landscape layout. Within the options there is also a watermark button which gives you the ability to print one of a selection of background "watermarks" indicating "Confidential", "Do not copy", etc. If this option is selected the background message will be printed on each sheet.

16.2.2 The margins tab

When you print your work Excel divides the printed paper area into several sections – the body of the worksheet, left and right margins, top and bottom margins, and header and footer sections. Most printers do not print within a quarter or half an inch of the edge of the paper, but Excel by default increases the margins as follows :

Left and right – 1.9 inches
Top and bottom, to header and footer – 1.3 inches
Top and bottom, to body of table – 2.5 inches.

You can adjust these margins by selecting the margins tab on the dialog box. To change the size of the margin, simply insert your desired value in the box.

16.2.3 The header and footer tab

You may find it useful to put the name of your workbook, or a suitable title, at the top of every page. To do this, select the down arrow next to the header text box. One of the options will be the name of your workbook. Select this. An example of what it will look like will appear in the sample box above. You might also like to put page numbers into your document. Click on the down arrow next to the footer dialog box and select page 1.

16.2.4 The sheet tab

Finally gridlines can be unattractive especially since Excel only prints the parts of the rows and columns used. To clear them from the printout, click on the sheet tab, and if the gridlines box is ticked then select it to remove the tick.

16.3 Print area

You may not want to print the whole of a worksheet, and this command enables you to select part only of a sheet. Drag over the area which you want to print and click on **File** ⇒ **Print Area** ⇒ **Set Print Area**.

16.4 Print preview

To view the effect on your work select the Print preview button on the dialog box. If you have more than one page you can flick back and forth through the pages using the next and previous buttons. To bring the page set-up dialog box back select the Set-up button at the top of the preview screen.

If you are still in Print Preview click OK when you have made all the desired changes on the page set-up dialog box. If you are happy with what you see click Close to get back to the editable page to add more data or click the Print button. If you are on the editable page click **Menu** ⇒ **File** ⇒ **Print**. Under the Print section make sure you have selected **Entire workbook** if you want to print all of your work. If you want more than one copy specify the number under copies at the top. A shortcut way of printing from the original editable page is to click the print button on the standard toolbar.

16.5 Print

Menu ⇒ **File** ⇒ **Print** is the last step in printing your work. Unlike the other options in Excel this does not have a standard dialogue box because this depends on the type of printer you are using. However most printers request the number of copies to be printed, and many give the option of draft or best quality.

Note that in addition to using the **Menu** ⇒ **File** ⇒ **Print** command you can also access the Print dialogue directly from the **Page Setup** and **Print Preview** windows.

16.6 Exporting to other applications

Documents and presentations using your Excel work can be created by using Word and Power-point respectively. This can be done by using the cut and paste buttons on the standard toolbar. These are some of the standard features available in the Microsoft Office suite.

First of all you must ensure the application that you are exporting from and the application you are exporting to are both running. You can switch between the two easily by clicking the buttons that bear their name on the task bar at the very bottom of your computer. Another way is to hold down the alt key and press **Tab** on the keyboard. Repeat pressing **Tab** until you reach the application that you are looking for.[27]

---

[27] On the Macintosh use the icon in the top right-hand corner to see all running applications, and select the one you want.

Select the information that you want in your report or presentation. Select **Menu ⇒ Edit ⇒ Copy**. Now switch to the application that you want to export to. Position the cursor where you want your data to appear and select **Menu ⇒ Edit ⇒ Paste**.

How the information will appear in the destination document will depend on the formatting already applied there. Consult a manual on that application if you want to reposition or reformat the imported data.

Chapter 17

Security

17.1 The problems

Your computer holds valuable data, or at least it will do when you have studied this volume and applied your new knowledge to your daily work. You may therefore consider it advisable to take security measures to ensure that the time you have spent in compiling your worksheets would not have been better used in tending your garden. Some of these, with their limitations, are:

(a) Lock your computer to the wall with security cable, fit security locks throughout your building, surround the building with a 10-foot steel fence and half-starved rotweilers, and install automatic guns. Others may not be able to gain access, but neither will you if you lose your keys or get eaten or shot on the way in. These measures will not prevent your computer experiencing a major system malfunction which wrecks your hard drive, or a lightning strike which fries the lot. Also, if you import any software or files from elsewhere, including the internet, any computer viruses which travel down the line will have a field day munching through your hard disk.

(b) Backup your files onto floppy disks or other media which can be removed from the vicinity of your computer and locked away. This will not prevent someone from switching on your computer and seeing and copying the files anyway, nor from computer viruses, which you will of course have automatically transferred to your backup system. Your computer may be given legs-assistance to disappear forever, but you can restore your files to the bigger and better machine which your insurance company will provide.

(c) Define a password for your computer. This is easily done and will prevent anyone else from starting it up and looking at your files. It will also stop you if you forget the password. It still does not avoid the virus problem or the Houdini-disappearance act.

(d) Define a password for each individual file. This can be effective in depriving all those who would like a freebie look at your confidential data files or your private letters to your mistress/master. You will need to apply a password to every file individually, and trying to remember the right password out of the 1,000 you have used will give you months of entertainment.

17.2 Possible solutions

Now for some serious suggestions. As indicated above, one measure is not sufficient for full protection, so you should consider a combination.

(a) Most computer cases have a security slot which enables the machine to be locked to the wall or substantial piece of furniture with a security cable or chain. Lock it up.

(b) Create a password for the computer. Some operating systems allow more than one password, and it is then possible to arrange for several people to use the machine, with each one only having access to their own files. Someone should be appointed as the Adminstrator, otherwise there may be conflicts. Go to **Menu** ⇒ **Tools** ⇒ **Options** ⇒ **Save** and click on **Always Create Backup Copy**. If your working copy is corrupted because of a machine crash or similar event there is a good chance that you can recover at least some of your work.

(c) Apply a password to important files. The safest procedure is to use just one or two different passwords, writing them down in a safe place. Change them on a regular basis. Applying a password will also encrypt the file so that data cannot be retrieved by other types of software.

(d) It is possible to control access either to a workbook or to individual parts of a worksheet or workbook. We may for example wish to allow others to have access to a worksheet in order to enter data, but not to see confidential information or formulae, or to change other visible information. Some of the options are:

(i) Deny access to anyone who does not know the password. To do this, use **Menu** ⇒ **File** ⇒ **Save As** ⇒ **Options**. You may now insert a password to open the file and another one to allow changes to be made. Your password can contain any combination of letters and numbers, other characters, but note particularly that passwords are case-sensitive – there is a difference between upper and lower case, and a common error is to enter a password when Caps Lock is on. Do not choose obvious words such as your partner's first name however fond you are of her/him, and enter your chosen password carefully as the characters do not appear in the box. You will be asked to repeat the password for safety reasons, and you should write it down as if you forget it there will be no way that you can ever open the file again.

(ii) You also have an option at this point to automatically save a backup which is a very useful safety precaution.

(iii) Protect formulae in a worksheet.. There are two stages in this. First select the cells containing the formulae and then **Menu** ⇒ **Format** ⇒ **Cells** ⇒ **Protection** and click on hidden. The second stage is to select **Tools** ⇒ **Protection** ⇒ **Protect Sheet**. The formulae will then be hidden, although the calculated result will still appear.

(iv) Unlike formulae, data cannot be hidden, but can be protected from alteration. Use the same procedure to format cells to be protected, and then use the protect sheet option as before. However it is possible to hide a complete column or row by formatting it as protected, dragging its width (or height) to zero, and using protect sheet to prevent the column or row from being opened up again.

(v) Make the whole or part, including formulae, read-only so that others can see data but not alter it. Proceed in the same way as for hiding formulae but select "Locked" only in the dialogue box.

(vi) The structure or windows of the entire workbook can be protected with **Menu** ⇒ **Tools** ⇒ **Protection** ⇒ **Protect Workbook**.

(vii) Protection is removed by **Menu** ⇒ **Tools** ⇒ **Protection** ⇒ **Unprotect Worksheet** (or Workbook if appropriate). The protected status of cells can then be cleared. Note that if a password had been used when protecting the workbook or worksheet re-entry of this is essential before alterations can be made, and as with using a password to open a file, forgetting the password is fatal – write it down and keep in a safe place.

(viii) Sometimes it is convenient to prevent certain parts of a worksheet from appearing in a printed copy even though they are on the screen. This can be done by selecting the cells and then **Menu** ⇒ **Format** ⇒ **Cells** ⇒ **Font**, and setting the colour to white.

(e) Save your important files onto floppy disks or removable disks and store them in a safe place away from the computer, or even better, at a different site. Use a rotation system with three or four disks so that even if corrupted files are saved you can still go back to an earlier version and hopefully retrieve most of your work. If you have an internet connection it may be possible to arrange with your Provider to upload files to the Server, which ensures that you have a copy off-site. However these files should be encrypted with a password so that they cannot be intercepted by anyone else.

(f) Viruses, worms and trojan horses are a serious menace. Viruses are little programs written by bright young computer enthusiasts, who cannot think of anything better to do with their expertise. These programs attach themselves to files and have the property that, just like medical viruses, they can replicate and attach copies of themselves to other files, particularly system files. What they actually do to your computer depends on the instructions which they carry. They may be designed to corrupt files every Friday 13th, put bouncing balls on the screen which wipe out other information, or just delete files from your hard drive.

(g) Worms are similar programs but are stand-alone. If your computer becomes infected they may eat away at your files, either deleting them, or just corrupting the contents.

(h) Trojan horses work differently. You run a new application which you have been given and which promises to be the solution to every problem you can imagine, or the most exciting version of *KlingOffs And The Martyrs* yet produced. While you are running this, the trojan horse, which is part of the software, is quietly digesting your most carefully constructed Excel workbooks.

(i) There are several programs which are designed to load and run in the background while you are doing your work. They will check for any virus-like activity and warn you, and if possible destroy the infection. You should look for a recommended program and read the instructions carefully.

Appendix 2

Glossary

Many of the following terms relate to Excel and have been discussed. They are included for reference. The remainder are terms which you are likely to come across at some time.

Application Another name for a computer program which performs complex operations, such as Excel.

Archive A file saved, usually on floppy disk, and stored elsewhere for safe keeping. Files may be combined or compressed for this purpose to save space.

ASCII The American Standard Code for Information Interchange. One Byte has 256 possible values, and each keyboard character, and many other codes, are assigned to one value. For example Capital A has a decimal value of 65 or a hexadecimal value of 41.

Background printing An arrangement by which, if you start printing, your work is "printed" to a new file, and then, while you continue working, the computer will send this file to the printer and delete it when printing is complete.

Backup The practice of saving a copy of a file for safety in case the working copy is corrupted. Excel can automatically save the previously saved version as a backup when doing a new save.

Baud, Baud rate and BPS The speed at which a modem operates. BPS (bits per second) is the actual rate of transfer, and Baud is approximately the same. A common bps rate is 38,400, which is a speed of 4,800 ASCII characters per second. Downloading large files from the Internet can take quite a long time!

Bit A Binary Digit, the smallest unit of information. It consists only of an on/off switch in the computer or a similar record on a disk.

Bitmap The image on the screen and on many printers is composed of many small dots, collectively known as a bitmap.

Boolean A form of algebra based on the work of the English mathematician, George Boole (1815–1864). It deals with the analysis of logic using a true/false basis represented by zero and one, and is the basis of all the operations which happen in your computer. (Boole was practically self-taught and had no degree, but became a Professor of Mathematics and a Fellow of the Royal Society.)

Boot blocks The sections of the startup disk which contain the information to start up your computer. When you switch on, a very small built-in program instructs the computer to look for the startup disk and load these blocks of information. This then instructs the computer to load the main operating system and other control programs.

Buffer An area of memory where data can be stored temporarily. Sometimes the computer may be busy organising itself while you are typing, and you may see a slight delay before characters appear on the screen. As you are typing your text it is being stored in a buffer until the computer is ready to deal with it.

Bug A malfunction of a computer program which gets its name from a mysterious malfunction in an early mainframe computer which was found to be due to a moth inside the machine. Bugs are likely to cause unexplained crashes of the computer.

Byte A set of eight bits – see ASCII.

Crash A situation where the computer suddenly stops operating for no apparent reason. Often the only solution is to restart it by pressing CNTRL ALT DELETE (PC) or CTRL COMMAND POWER (Macintosh) simultaneously, when all information will be lost. One cause of crashes is lack of memory when two applications try to use the same area. Possible solution – install some more memory.

Cursor The marker which appears on the screen at the point where the next character will be entered.

Database A file in which data records are stored in a form in which they can be retrieved easily. An Excel file can be used as a database.

Data file Any file in which data are stored, including word processing documents. Other classes of file are applications and utilities.

Debugging The process of correcting errors in an application. Modern applications are so large that it is impossible to be sure that there are no errors, and problems are often found after they have gone on sale. Software manufacturers frequently issue modifications to correct these problems.

Directory (PC) A disk structure which holds several files. Each directory may contain other directories, and thus a tree structure can be created, allowing files to be classified in much the same way as a library and retrieved easily.

Field In a database, a unit of information about a record, such as a postcode. When using Excel as a database, a column can be designated as a field for a particular type of data such as the postcode.

File format Files can be written to disk coded in different ways, depending on the objective. A normal save in Excel uses a format specially designed for efficiency, but this cannot usually be read by a different application. If however it is saved using, for example the SYLK format, another spreadsheet application, or even an earlier version of Excel, will be able to read it. Be aware however that in this process some cell formatting or other characteristics may be lost.

Floppy disk A thin piece of plastic coated with a magnetic surface on which data can be written. All modern disks are encased in a plastic cover for protection.

Folder (Macintosh) A disk structure which holds several files. Each folder may contain other folders, and thus a tree structure can be created, allowing files to be classified in much the same way as a library and retrieved easily.

Font Software that creates a particular typeface on the screen or printout. The term is often used also to refer to the typeface itself.

Footer A section at the bottom of a printed page in which you can insert a page number, date, time, reference, or any other information which will appear on each page.

Formatting (1) Another name for initialising a disk. (2) Defining the size, font and style of text, the background, and other features related to the appearance of your work.

Fragmenting When a disk has been in use for some time, files have been deleted, and new files saved, it will eventually happen that files are split and saved on different parts of the disk. This will slow down the system because it takes more time for the computer to find the various parts of the file. There are special utilities designed to defragment disks.

Function keys The 12 (or 15) keys at the top of the keyboard which can be designated to perform special operations, depending on the application in use.

Gigabyte One thousand million bytes.

Handshaking The process by which two connected computers will initially agree on the way in which they will transfer data, including the speed at which they will do it. You don't see this operation and don't need to know anything about it if it works correctly.

Hard disk (or Hard drive) A storage device consisting of one or more metal plates coated with a magnetic surface. It will normally contain the software for starting the computer in addition to all the applications and data files.

Hardware The physical parts of a computer, memory chips, power supply, monitor, etc. Originally so-called because early computers were very large (and slow) and looked more like a hardware store than a machine for doing calculations.

Header Similar to a footer, but at the top of the page.

Hertz Cycles per second. Used in particular to measure the speed of the internal clock which these days is measured in Megahertz (million cycles) per second.

Hexadecimal An alternative numbering system used in computing and operating on Base 16 instead of our usual Base 10 (decimal) system. The digits are 0,1,2,3,4,5,6,7,8,9,A,B,C,D,E,F, and this is much more convenient in machine code calculations than decimal because each ASCII code can be designated by just two digits.

I/O Input/output. The means by which the computer is connected to other devices such as printers, modems and scanners. Connection is by cables plugged into the sockets on the back of the computer.

Initialize Preparation of a disk for storage of programs or data. When new, a disk has no magnetic markings and there would be no means of recording the position of data. The initialisation process marks the disk into concentric tracks and divides each track into a number of sectors. Each sector is numbered and part of the disk is reserved for the recording the address where each file is stored, and sectors which are empty.

Inkjet A very common type of printer which sprays small droplets of ink on to the paper (about 300 dots per inch on each line) to form characters.

K Literally, an abbreviation for 1000 (compare kilometres)

Kilobyte One thousand bytes. Each byte holds one alphabetic character, but an Excel workbook will use many more bytes to hold formatting information, formulae, etc.

Laser printer A printer which uses a laser to create an electrostatic image of the output on a special drum. This drum then transfers the black toner to the paper. More expensive, but better quality than an inkjet printer.

LCD Liquid Crystal Display of the type used on laptop computers.

Lock A means of preventing something from being altered. Files can be locked so that they cannot be changed. In Excel individual cells can similarly be locked with **Format** ⇒ **Cells** ⇒ **Protection**, but this will not actually have effect until the worksheet itself is protected with **Tools** ⇒ **Protection** ⇒ **Protect Worksheet**. A password is optional.

Macro A series of instructions which can be set up so as to be carried out by a single keystroke.

Megabyte One million bytes. A standard high-density floppy disk holds about 1.4 megabytes.

Megahertz Million cycles per second.

Memory The part of the computer which holds the data on which you are working, and the applications which you are using. These days you should expect to have at least 32 megabytes, and more if you intend to do graphics work.

Modifier key One of a group which includes Shift, Caps Lock, Control, Alt (PC) and Option and Command (Macintosh). Modifier keys generally do not perform any function by themselves but alter the effect of other keys.

Nanosecond One billionth of a second. It is commonly used to specify the access time to a byte of memory, for example 80 ns. This is one of the critical factors in determining the overall speed of the computer.

OCR Optical character recognition. A useful means of transferring data to a worksheet from a printed page is to scan it and convert it directly to an Excel file.

Operating system The software which makes the computer work.

Parallel port Port for sending data along eight lines simultaneously. A whole byte of data is therefore sent at one go. (C/f serial port)

Pixel A dot on the computer screen. A typical screen is composed of 640 x 480 dots on the same principle as a television set but designed to give a much higher definition.

Platform The type of operating system such as Windows 95, Windows NT or Macintosh. A cross-platform application can write files in the format required by another operating system.

Port A socket on the back of the computer into which a cable from a peripheral such as a printer can be plugged.

PostScript A definition of characters or other information for printing by formula rather than by a pattern of dots. Laser printers normally use PostScript which gives a much higher definition.

Processor The "brain" of the computer. It reads instructions from the memory and carries them out. For example, it may receive an instruction from Excel to get the number from A1, add the number in A2, and store the result in A3. It also carries out instructions relating to system operations.

RAM disk This is a facility which allows you to designate part of the memory as a temporary disk. The advantage of this is that frequently-used parts of the program can be transferred to this area and access is then much quicker than reading those parts from the disk. Assuming you have the facility, once you have turned it on you don't need to know anything more about it. Note however that some applications do not like it.

Record The information relating to each case in a database.

ROM Read-only memory. This is memory permanently stored within some of the chips in the computer and which controls some of the basic operating functions.

Scanner A device for converting a printed page into digital form and transferring it to the computer for processing.

Screensaver If a monitor is left to show the same output for a long period (usually several hours) the image may become burnt into the screen and not removable. A screensaver is a program which cuts in after a predetermined time (usually a few minutes) and blanks the screen or shows a constantly varying pattern, sometimes with built-in humour. Its use is recommended.

Serial port A port for passing data along a single conductor in a serial cable. Bits are sent one at a time to the receiver.

Software The application programs which instruct the computer on its mission in life. Some of these are permanently built into the computer and cannot be altered. Others are loaded at startup or during operations, but are not obvious to the operator. The most obvious ones are those loaded into the computer by the operator, such as Excel. (C/f Hardware).

SYLK Symbolic Link format. A format for worksheet files which enables them to be read by other spreadsheet applications.

Typeface The style of the characters displayed or printed by the computer, such as Times, Helvetica or Old English. There is a huge variety of typefaces now because designing them has become relatively easy.

Utility A program, usually fairly small, which gives support to operations. Examples are virus seekers and defragmenters.

Virus A nasty little program designed by clever boffins who have nothing better to do with their time. It attaches itself to other files in your computer and the replicates itself. It may proceed to destroy other files or perhaps cause your machine to crash on particular days.

WYSIWYG ("wizzywig") "What you see is what you get". Early desktop computers would re-format pages of output before printing so that you did not know how it would look until it was actually on paper. Today practically all programs print your efforts at word processing or compiling worksheets exactly as you see them on the screen.